Ecology and Behaviour of Nocturnal Primates

Ecology and Behaviour of Nocturnal Primates

Prosimians of Equatorial West Africa

Pierre Charles-Dominique

Translated by R.D. Martin

New York Columbia University Press 1977

First published in 1977 in Great Britain by Gerald Duckworth & Co Ltd
and in the United States of America by Columbia University Press

Printed in Great Britain

Library of Congress Cataloging in Publication Data

Charles-Dominique, Pierre.
    Ecology and behaviour of nocturnal primates.

    Bibliography: p.
    Includes index.
    1. Lorisidae – Behaviour. 2. Nocturnal animals –
Behaviour. 3. Mammals – Behaviour. 4. Mammals – Ecology.
5. Mammals – Gabon. I. Title.
QL737.P955C4513   1977   599′.81   77-1227
ISBN 0-231-04362-7

Contents

Author's Preface

The field studies outlined in this book were carried out at the Laboratoire de Primatologie et d'Ecologie des Forêts Equatoriales (supported by the Centre National de Recherche Scientifique) and depended upon the kind hospitality of the government authorities of Gabon and upon the valuable collaboration of the laboratory staff, past and present.

Special thanks are due to my friend and colleague Dr. R.D. Martin for accepting the task of translating the manuscript and for his advice and comments in the preparation of the text. Thanks go to my wife, Mireille Charles-Dominique, and to Catherine Muñoz-Guevas for preparing the line-drawings, to Jeremy Dahl for preparing the index, and also to Alain R. Devez for providing many of the photographs.

Finally, I would like to thank my publishers for their personal interest and guidance.

This is the first publication of the book in any language.

P.C.-D.

Translator's Preface

Human beings are accustomed to daytime activity and to relying predominantly upon their well-developed sense of vision for dealing with their surroundings. Hearing and olfaction are relatively far less important, and human beings can thus most easily understand the behaviour of animals which are similarly diurnal in habits. Yet the majority of living mammal species are nocturnal in habits, and there are many lines of evidence indicating that the ancestors of the marsupial and placental mammals were adapted for nocturnal life. In a recent review, Pierre Charles-Dominique has himself shown, from an analysis of modern faunal compositions in tropical rainforest areas of West Africa and South America, that birds occupy most of the available ecological niches during the daytime. With few exceptions, mammals within the range of body sizes covered by the birds are nocturnal or crepuscular in habits, feeding in darkness on food types which are primarily exploited by birds during the daytime. It is only the larger-bodied mammal species which have become adapted in significant numbers to diurnal life, and the shift from nocturnal habits has invariably been accompanied by specialisations, notably in the visual apparatus and the central nervous system.

The primates themselves show a sharp demarcation between the more primitive 'prosimians' and the highly specialised 'simians', including man. The prosimians are predominantly nocturnal and relatively small-bodied, whereas the simians are diurnal, with only one exception (*Aotus trivirgatus*), and their average body size is considerably greater. Thus, the shift from nocturnal to diurnal life is one of the major factors to be considered in reconstructing the evolution of the primates. We therefore need to understand both nocturnal and diurnal primate species.

Study of the social behaviour of individual primate species, and especially of the relationship of behaviour to ecology, is one of the most promising current areas of research relating to human origins. Further, detailed analysis of the links between behavioural adaptations and specific ecological conditions ('behavioural ecology') is an important field in its own right, and one which is receiving increasingly more attention in field studies. However, until about fifteen years ago field studies of primates were largely concerned with diurnally active simian species, and even the first studies on prosimian species were largely devoted to the few diurnal lemur species surviving on Madagascar. Jean-Jaques Petter, who pioneered much of the field work on the lemurs as a group, in fact conducted some of the earliest coordinated studies on the nocturnal lemur species, and his doctoral thesis (published in 1962) is a landmark in the development of field research on prosimians. He made extensive use of the head-lamp technique of locating lemurs at night by searching for reflections from their eyes. The tapetum behind the retina of each eye in nocturnal lemurs, which represents an adaptation to assist vision at low light intensities, gives a bright orange-yellow reflection in the light of the lamp. Observations conducted in the forests at night using this technique for location of nocturnal lemurs permitted Petter to record new information which laid much of the basis for later comparative study of the behaviour and ecology of nocturnal primates. In a similar way, Pierre Charles-Dominique's study of individually marked prosimians, belonging to five different species living in the same area, represents a landmark. His meticulous observations in the equatorial rainforest of Gabon, extending over the past decade, have involved trapping, marking and release, direct observations, experimental interventions under natural conditions, and – very recently – the application of radio-tracking techniques. It would be no exaggeration to say that the major part of available information about the natural behaviour and ecology of the African prosimians (bushbabies, pottos and angwantibos) has derived from his work. For each of the five species in Gabon, he has built up a detailed picture of diets, locomotion, defence against predators, patterns of activity, sleeping sites, population parameters and social behaviour, through a combination of direct and indirect observations conducted under natural conditions. For each species, therefore, one can see how behavioural adaptations relate to local ecological conditions.

The importance of Pierre Charles-Dominique's studies goes far beyond the simple description of behaviour and related ecological factors for the five prosimian species. His contribution to behavioural ecology is of special value because the five prosimian species studied all occur together and can potentially enter into competition for food resources and space. Careful analysis of the ways in which the five species avoid competition throws new light on one of the major evolutionary processes leading to increased specialisation. Each prosimian species in Gabon has a well-defined ecological niche. First of all, there is a sharp demarcation between the three bushbaby species (Demidoff's bushbaby; Allen's bushbaby; needle-clawed bushbaby) and the two lorisine species (potto;

1

Introduction

It is a common misconception that the term 'primate' refers only to the apes or monkeys. Indeed, only one or two dozen species of the Order Primates are well known to non-specialists, whereas in fact there are about 188 living species (if the tree-shrews are excluded). Living primate species are generally distinguished from placental mammals belonging to various other Orders by the following notable features: the hands and feet are adapted for grasping, in association with the typically arboreal way of life; the sense of vision is predominant, with forward rotation of the eyes for emphasis on binocular sight accompanied by increased complexity of the visual apparatus; and the brain is relatively enlarged as a correlate of increased behavioural flexibility and complexity. Man and the so-called 'higher' primates (apes and monkeys) fit this definition quite well, but as one passes back through the fossil record towards the ancestral primate stock it becomes increasingly difficult to apply the same criteria. However, the same problem applies to all mammalian Orders, and near the origin of any evolutionary tree it is difficult to decide upon the classification of certain little-differentiated fossil forms. In fact, such a decision is largely arbitrary and the important thing is to recognise the 'evolutionary pathways' followed by the various placental mammal groups. In the case of the Order Primates, we are concerned with the evolutionary developments which have taken place within the group to which we ourselves belong. The study of evolution is not, in itself, the only goal of primatologists; but such study greatly contributes to better understanding and interpretation of numerous aspects of primate biology.

In most current classifications (e.g. Simpson, 1945) the primates are divided into two main groups, the prosimians (so-called 'lower primates') and the simians (so-called 'higher primates'), as is shown in Table 1. The simians are divided into two geographical groups which have followed very similar evolutionary pathways, the New World monkeys (Superfamily Ceboidea) and the Old World monkeys, apes and man (Superfamilies Cercopithecoidea + Hominoidea). These two large groups together contain 70% of the living primate species, whereas the surviving prosimians represent only 30%.

Table 1 Classification of living primates (following Simpson 1945, but excluding the tree-shrews); common names in brackets

Order : PRIMATES (primates)
    Suborder : PROSIMII (prosimians)
        Infraorder : LEMURIFORMES (Madagascar lemurs)
            Superfamily : LEMUROIDEA
                Family : 1. LEMURIDAE
                    Subfamily : 1. LEMURINAE (true lemurs)
                            2. CHEIROGALEINAE (mouse & dwarf lemurs)
                Family : 2. INDRIDAE (indri group)
            Superfamily : DAUBENTONIOIDEA
                Family : DAUBENTONIIDAE (aye-aye)
        Infraorder : LORISIFORMES
                Family : LORISIDAE (Afro-Asian lorisids)
                    Subfamily : 1. LORISINAE (lorisines)
                            2. GALAGINAE (bush-babies)
        Infraorder : TARSIIFORMES (tarsiers)
                Family : TARSIIDAE
    Suborder : ANTHROPOIDEA (simians)
            Superfamily : CEBOIDEA (New World monkeys)
                Family : 1. CEBIDAE
                    2. CALLITRICHIDAE
            Superfamily : CERCOPITHECOIDEA (Old World monkeys)
                Family : CERCOPITHECIDAE
            Superfamily : HOMINOIDEA (apes and man)
                Family : PONGIDAE
                Family : HOMINIDAE

The first fossil forms identified as simians appear only 30-35 million years ago (Oligocene primates of the Fayum), whereas fossil forms regarded as prosimians can be traced back to more than 60 million years ago (Cretaceous/Palaeocene boundary). The early prosimians gave rise to lineages leading to the three surviving prosimian groups (Madagascar lemurs; Afro-Asian lorisids; tarsiers) and to the ancestors of the living simians. It has been suggested that the simian primates were more successful in competition and largely replaced the prosimians on the major continental areas, leaving only a small number of nocturnal forms occupying narrow and specialised ecological niches not exploited by the monkeys or by other mammal groups. This would apply to the South-East Asian tarsiers (3 species) and the Afro-Asian lorisids (11 species). The Madagascar lemurs, isolated from the simians now widespread on the African mainland, underwent independent adaptive radiation to give rise to large-bodied diurnal forms (10 surviving species and approximately 10 subfossil species presumed to have been diurnal in habits) alongside various other small-bodied nocturnal forms (11 surviving species). Among the living simian primates, only the night monkey (*Aotus trivirgatus*) of South America is nocturnal and is assumed to be secondarily adapted for this mode of life (Hill, 1953; Le Gros Clark, 1971).

As a general rule it would seem to be the case that diurnal life presented great evolutionary possibilities for the primates, whereas nocturnal life blocked or retarded their evolution. Thus, the nocturnal primates represent a more primitive condition based on the retention of numerous little-modified ancestral features, combined with a relatively small number of recent specialisations. For this reason, study of the biology of prosimians is of capital importance for the recognition and interpretation of the evolutionary history of the primates, including that of man.

It is customary to classify the lorisids and lemurs in two separate high-level categories (e.g. Infraorders), for example as Lorisiformes and Lemuriformes (Simpson, 1945: see Table 1) or as Lorisoidea (Tate Regan, 1930) and Lemuroidea (Mivart, 1864 restricted). In addition, some authors (e.g. Hill,

1953) group the lemurs and lorises together in the grade Strepsirhini E. Geoffr. In fact, the justification for separate high-level classification of the lemurs and lorises is largely derived from comparisons conducted between anatomical characters of certain slow-moving lorisids (*Perodicticus* and *Loris*) and of certain large-bodied lemurs (particularly *Lemur*), as is evident from the publications of van Kampen (1905), Pocock (1918), Weber (1928), Le Gros Clark (1971) and Hill (1953). Yet comparison of the morphologically least specialised nocturnal members of the two groups, the Galaginae (bushbabies) and the Cheirogaleinae (mouse and dwarf lemurs), reveals a considerable number of similarities, which are apparently largely based on retention of numerous common primitive characters from the ancestral primate stock (Charles-Dominique & Martin, 1972b; see Conclusions). The taxonomic status of the Lorisiformes and the Lemuriformes will perhaps be subject to revision when more is known about the biology of these prosimians. Nevertheless, it is probably true that the two groups have been partially or completely separated for a long period of time by the Mozambique Channel, which had formed by the beginning of the Tertiary and has apparently continued to widen since that time (Martin, 1972). Effective biological isolation of Madagascar is doubtless more recent than the earliest physical isolation, but unfortunately no relevant early Tertiary fossils have been found as yet to indicate actual dates. On the other hand, Miocene sites in East Africa have yielded definite lorisid material, which shows that this group had become established by at least 20 million years ago. These Miocene fossil lorisids, most of which seem to be related to modern Galaginae rather than Lorisinae, indicate a size range comparable to that found among living bushbabies, and there would seem to be little morphological distinction apart from the lesser development of the tarsal area of the hindlimb in the fossil species (Walker, 1974).

The living Lorisiformes can be sharply divided into two groups usually given the rank of subfamily (Mivart, 1864), but occasionally recognised as two separate families (Weber, 1928). The following detailed classification has been adopted for use in the text (see also Table 1).

Family: Lorisidae (Afro-Asian lorisids)

Subfamily 1: Lorisinae (lorisines). Slow-moving animals, with a greatly reduced tail and atrophy of the second digit of the hand. They are exclusive climbers and never leap. There are two Asiatic species, the slender loris (*Loris tardigradus*) and the slow loris (*Nycticebus coucang*);[1] and two African species, the potto (*Perodicticus potto* – Fig.1) and the angwantibo (*Arctocebus calabarensis* – Fig.2).

The lorisines inhabit dense forests, usually in zones where lianes are abundant. They have not succeeded in colonising relatively 'open' habitats in more arid regions, as has been the case with the bushbabies (galagines), which can leap between trees separated by gaps.

Subfamily 2: Galaginae (galagines). Fast-moving animals with a long tail and powerfully developed, elongated hind limbs. All six recognised species are confined to Africa. The Senegal bushbaby (*Galago senegalensis*), which occurs to the north, east and south of the Congo and Guinea rain-forest block, and the thick-tailed bushbaby (*Galago crassicaudatus*), which occurs to the east and the south of the same forest block, both inhabit bush, wooded savannah or gallery forest, where they may be sympatric (Bearder and Doyle, 1974). These two species (Fig.3) probably originated in dense forest areas, subsequently adapting to drier conditions peripheral to the rain-forest (Walker, 1974). The other three main species, Demidoff's bushbaby (*Galago demidovii* – Fig.4), Allen's bushbaby (*Galago alleni* – Fig.5) and the needle-clawed bushbaby (*Euoticus elegantulus* – Fig.6), occur in the Congo and Guinea rain-forest block along with *Perodicticus potto* and *Arctocebus calabarensis*. The sixth species, *Galago inustus*, was initially considered to be a subspecies of *Galago senegalensis* (*G.s. inustus* Schwartz 1930), then as a species of the genus *Euoticus* (*E. inustus*) by Hill (1953) (following data from Hayman, 1937), before its classification by Walker (1974) as a species of the genus *Galago*. This bushbaby species is apparently confined to a zone covering the east of Zaire and the west of Uganda (Vincent, 1972).

---

[1] Tien (1960) regards the slow loris of Vietnam as a separate species – *Nycticebus pygmaeus*.

Figure 1. The potto, *Perodicticus potto edwardsi* Bouvier, 1879. Vernacular names: Fang – *Awoun*; Bakota – *Ikoumama*; Bakwélé – *Dira*. (Photo: A.R. Devez)

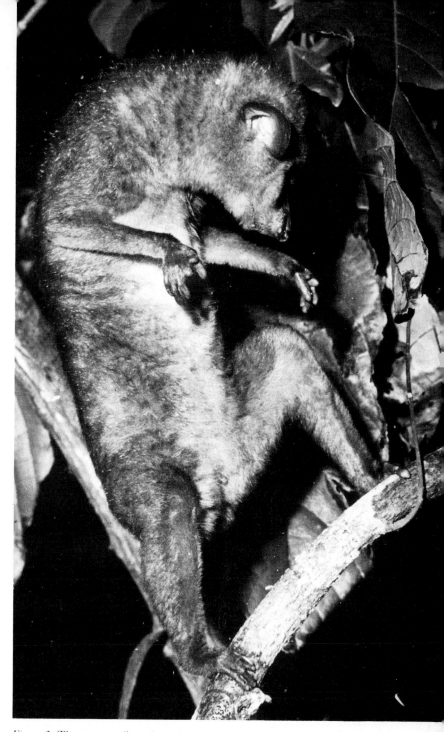

Figure 2. The angwantibo, *Arctocebus calabarensis aureus* (De Winton, 1902) Vernacular names: Fang – *Mese Tsere*; Bakota – *Koundé*; Bakwélé – *Sologou*. (Photo: A.R. Devez)

Figure 3 A and B.
A. The Senegal bushbaby, *Galago senegalensis moholi*, in its natural habitat: *Acacia* wooded savanna in South Africa. (Photo: R.D. Martin)

B. The thick-tailed bushbaby, *Galago crassicaudatus umbrosus*, in its natural habitat: dense evergreen forest and riparian bush in the Transvaal, South Africa. (Photo: S.K. Bearder)

Figure 4. Demidoff's bushbaby, *Galago demidovii* (Fischer, 1806). Vernacular names: Fang – *Odzam*; Bakota – *Ndè*; Bakwélé – *Sèb*. (Photo: A.R. Devez)

Figure 6. Euoticus elegantulus elegantulus (Le Conte, 1857), the needle-clawed bushbaby. Vernacular names: Fang – *N'sé*; Bakota – *Ibobo*; Bakwélé – *Egûel*. (Photo: A.R. Devez)

Figure 5 (left). Allen's bushbaby, *Galago alleni* Waterhouse, 1837. Vernacular names: Fang – *Emmam*; Bakota – *N'gokoué*; Bakwélé – *N'gok*. (Photo: A.R. Devez)

The five lorisid species which occur in the large equatorial rain-forest block (3 galagines and 2 lorisines) are sympatric in Nigeria, Cameroon, Rio Muni, Gabon, Central Congo and the Central African Republic; but the geographical ranges of two of them (*Galago demidovii* and *Perodicticus potto*) extend farther to the north (into Senegal) and to the east (up to the Rift Valley). The broad patterns of geographical distribution of the various lorisid species are given in Fig.7, while Table 2 lists their body weights and physical dimensions.

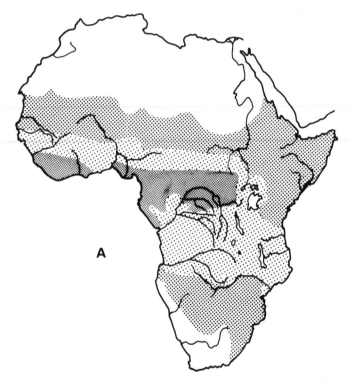

Figure 7 A to D. Geographical distribution of the various lorisid species (following Hill, 1953 and Vincent, 1969).

A. Vegetational zones of Africa, following the UNESCO *Carte de la végétation de l'Afrique* (Aubreville, A.A., Duvigneau, P., Hoyle, A.C., Keay, R.W.J., Mendonca, F.A., and Pichi-Sermolli, R.E.G., 1958).
central stippled area: dense rain-forest of low and medium altitude.
intervening stippled area: mosaic savannah forest, wooded savannah, herbaceous savannah.
peripheral stippled area: steppe grassland, subdesert steppe and mountain communities.

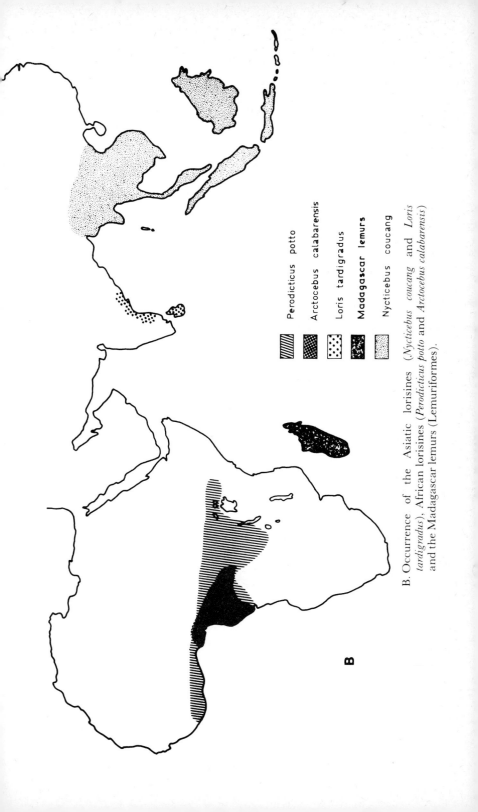

B. Occurrence of the Asiatic lorisines (*Nycticebus coucang* and *Loris tardigradus*), African lorisines (*Perodicticus potto* and *Arctocebus calabarensis*) and the Madagascar lemurs (Lemuriformes).

Perodicticus potto

Arctocebus calabarensis

Loris tardigradus

Madagascar lemurs

Nycticebus coucang

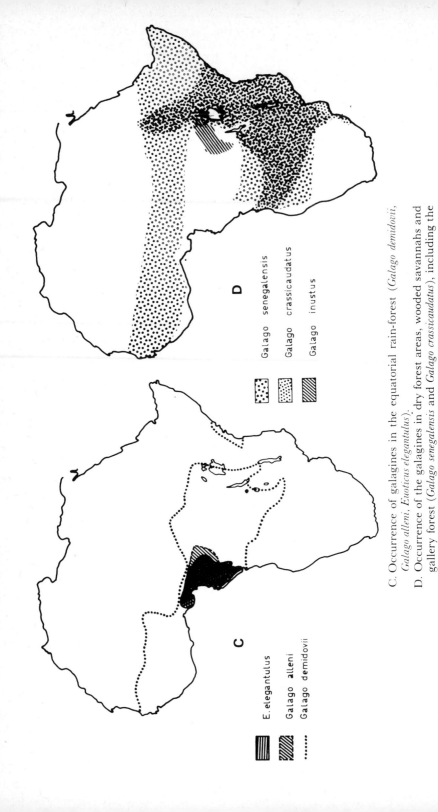

C. Occurrence of galagines in the equatorial rain-forest (*Galago demidovii*, *Galago alleni*, *Euoticus elegantulus*).

D. Occurrence of the galagines in dry forest areas, wooded savannahs and gallery forest (*Galago senegalensis* and *Galago crassicaudatus*), including the distribution of *Galago inustus*.

Galago senegalensis
Galago crassicaudatus
Galago inustus

D

E. elegantulus
Galago alleni
Galago demidovii

C

Table 2 Weights and measurements of African and Asian lorisids

	Body weight (gm)	Head and body length (range in mm)	Tail length (mm)
Galago crassicaudatus (1)	1000-2000	297-373	415-473
Galago senegalensis (2)	230-300	150-173	205-250
Galago inustus (3)	similar to *G. senegalensis*		
Galago demidovii N = 66 (4)	61 (46-88)	123 (105-123)	172 (150-205)
Galago alleni N = 17 (5)	260 (188-340)	200 (185-205)	255 (230-280)
Euoticus elegantulus elegantulus N = 39 (6)	300 (270-360)	200 (182-210)	290 (280-310)
Perodicticus potto edwardsi N = 33 (7)	1100 (850-1600)	327 (305-370)	50 (37-70)
Arctocebus calabarensis aureus N = 30 (8)	210 (150-270)	244 (230-260)	15
Nycticebus coucang (9)	1012-1675	265-380	—
Loris tardigradus (10)	270-348	186-264	—

Notes:

(1), (2), (9) and (10) following Napier and Napier, 1967; there are considerable subspecific differences.

(4), (5), (6), (7) and (8) represent data derived from adult males and females in the author's study area. Other subspecies may differ considerably (e.g. for *Arctocebus calabarensis calabarensis* the following figures are given by Napier and Napier, 1967: weight – 266-465 gm; head and body length – 220-603 mm; tail length – approx. 8 mm.).

(9): *Nycticebus pygamaeus* is considered as a species separate from *N. coucang* (Tien, 1960). Napier and Napier, 1967, give the head and body length of this species as 190 mm.

A. STUDY AREA

The field investigations on which this book is based were conducted in the Ogoué-Ivindo region of West Africa (Gabon), included in the Congolese rain-forest block (Fig.8). This region is still too remote from the transport network to be exploited by forestry companies, and the human population density is very low (less than two inhabitants per square kilometre). The villages and plantations are localised along rivers and roads in such a way that there are still vast zones of relatively accessible primary forest present. Access has been improved recently by the construction of roads in previously uninhabited areas and by the introduction of a new laboratory in a primary forest zone (Ipassa plateau). In this practically intact area of forest, comparative study of the ecology and behaviour of the five lorisid species is of great interest, since the habitat is probably similar to that in which they evolved and acquired their various specialisations.

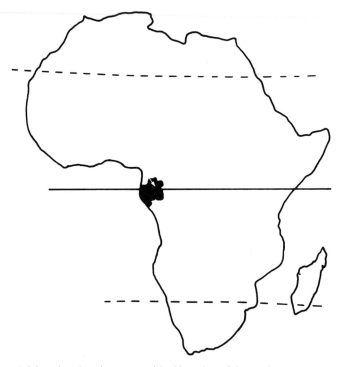

Figure 8. Map showing the geographical location of the study zone.

The overall field study of these primates was conducted between October 1965 and July 1973, involving seven visits covering a total of 42 months. Complementary investigations have been carried out in captivity with a colony of the five lorisid species, maintained at the Laboratoire d'Ecologie Générale of the Muséum National d'Histoire Naturelle in Brunoy (France). The base for the field study was the laboratory at Makokou, situated in the north-east sector of Gabon (0.4° latitude north), 500 kilometres from the Atlantic coast. The laboratory was originally founded by Professor P.P. Grassé in 1962 under the name of the 'Mission Biologique au Gabon', and it has now become the 'Laboratoire de Primatologie et d'Ecologie des Forêts Equatoriales', supported by the Centre National de Recherche Scientifique (CNRS) and directed by Dr. A. Brosset. The terrain surrounding the laboratory is relatively flat (average altitude: 500m), but there is considerable subdivision by valleys and water-courses. The forest covers almost all of the available surface, and even areas subjected to repeated flooding support forest vegetation, except in a very few, rare places where the ground is flooded continuously. The first occurrence of savannah is found at 300 kilometres to the south-east of Makokou. Around the villages, there are a number of zones deforested for agriculture; but the forest regenerates quite rapidly, with the result that all areas of human habitation and exploitation display the various stages of secondary forest side by side. Fig.9 shows the general pattern of forest around the laboratory.

B. CLIMATE

Climatic variation in the Makokou area, which is situated virtually on the equator, is essentially dependent upon rainfall (Fig.10). At Makokou, one can distinguish the following four seasons: the long rains, lasting from the end of September to the end of December; a short dry season from the end of December to the end of March; the short rains, lasting from the end of March to the end of June; and a long dry season from the end of June to the end of September. The average annual rainfall is 1,700 mm, spread over 100-120 days of the year.

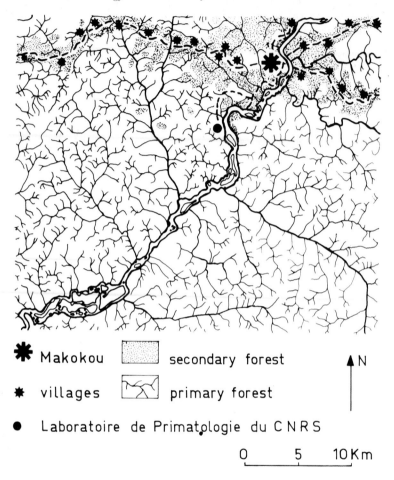

✳ Makokou ▨ **secondary forest**   ▲N

✳ villages ▨ **primary forest**

● Laboratoire de Primatɒlogie du CNRS

        0   5   10Km

Figure 9. The study area near to Makokou. Stippling shows secondary forest and cultivated areas adjacent to the road (dashed line). (Adapted from the map supplied by the Institut Géographique National, Paris)

Figure 10 (right). Annual variation in biomass of insects, fruits and flowers in relationship to climatic variation (data collected 1968). The long dry season occurs in July/August/September, and productivity reaches a minimum. The weights of insects are monthly means for daily captures made with an ultra-violet light-trap, illuminated for the first 4 hours of each night. Indices of abundance of fruits and flowers are based on counts of trees and lianes in fruit or flower along a standard transect of 3 km in primary forest, utilised on a regular basis.

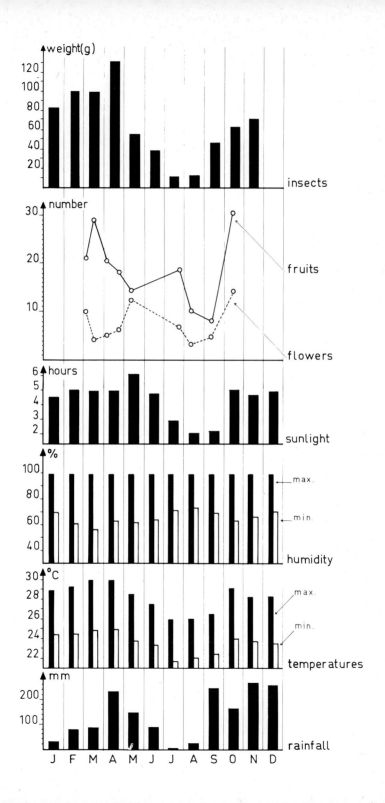

Figure 11. Aerial photograph of the primary forest in the Makokou region. The new CNRS 'Laboratoire de Primatologie et d'Ecologie des Forêts Equatoriales' is located in the clearing. Note the mist above the canopy. (Photo: A.R. Devez)

C. VEGETATION

In the primary equatorial rain-forest (Fig.11) one can distinguish three classical strata of vegetation, which intergrade quite smoothly (Fig.12). The largest trees reach a height of 50m and they provide points of attachment for numerous liane species which play an important part in the lives of climbing mammals by providing locomotor routes from one tree to another. The foliage of these lianes usually develops with full exposure to light, in the upper stratum of the forest, and it is often intimately entangled with the leaves of the supporting trees (Fig.13).

In the undergrowth, where light penetrates only as small, isolated patches, growth of vegetation is slow. The bushes have long, thin trunks and branches, and their foliage is relatively sparse. In the zone between ground-level and a height of 5 metres, the tree-trunks and liane bases constitute the bulk of the vegetation, and this permits relatively easy movement under conditions of fair visibility (Fig.14).

Figure 12(right). Direct view of the primary forest obtained following a recent clearing operation. (Photo: A.R. Devez)

Figure 13. 'Liane curtain', showing the details of a tree invaded by lianes in the canopy of primary forest. (Photo: A.R. Devez)

Figure 14. View of the undergrowth in primary forest. (Photo: A.R. Devez)

Trees fall quite frequently, particularly during tornadoes. The largest trees often drag down with them other, smaller trees and clumps of lianes, producing large clearings in the forest. The dead wood is rapidly exploited by termites, with the result that the sun soon penetrates directly to the ground after a tree has fallen and very dense vegetation, rich in lianes, subsequently grows up. Primary forest therefore exhibits great variation in composition, with clearings showing all stages of forest regeneration. Such heterogeneity and the associated complexity of the vegetation provide the basis for a highly diversified fauna exploiting the wide variety of ecological niches presented by the forest.

The Gabonese villagers generally follow a slash-and-burn tradition of agriculture, cutting down trees with axes before burning off the vegetation. Subsequent plantations of maniok, bananas and maize exhaust the soil quite rapidly, and cultivated areas are soon abandoned. Such previously cultivated areas are quickly colonised by various bushes and herbaceous plants, and numerous plant species appear and disappear in succession as the community matures. The parasol tree (*Musanga cecropioides*) is the most typical tree in such plant formations and it dominates the undergrowth, which is still dense and rich in lianes even after several decades (Fig.15). In the shade of this vegetation the seeds of the constituents of the climax forest undergo germination, slowly reconstituting the old forest.

In the Makokou region, plantations were still small-scale undertakings only fifty years ago, and the Gabonese villagers at that time lived predominantly by hunting, fishing and gathering. At the present time, secondary forests are largely restricted to the regions adjacent to communication routes, and new agricultural areas are usually established in relatively young secondary forest zones. A large number of mammal and bird species of the primary forest have colonised this secondary forest habitat, which is also inhabited by various species peculiar to secondary forest. Four of the five lorisid species of the primary forest are also to be found in secondary forest formations, and their ecological localisation within this recently established habitat, when compared with their original localisation in primary forest, often permits more detailed understanding of the exact nature of the biotope sought out by each species.

Figure 15. Aerial photograph of secondary forest. (Note the large numbers of parasol trees.) At the edge of the forest is the former CNRS 'Laboratoire: Mission Biologique au Gabon'. (Photo: A.R. Devez)

D. THE ECOSYSTEM

In order to understand fully the biology and evolution of the populations of the various lorisids, it is essential to examine them in terms of the overall equatorial forest ecosystem. In the Makokou region alone research workers have identified 120 mammal species, including 17 primates, and 216 bird species, while at the present time it is still impossible to make a comprehensive inventory of the lower vertebrate and invertebrate species. 80% of the mammals are nocturnal (some being both nocturnal and diurnal in habits), while 96% of the birds are diurnal. In many cases, birds and mammals exploit successively the same food-sources, the former by day and the latter by night. Among the nocturnal forest mammals there are 11 insectivores, 27 bats (7 megachiropterans and 20 microchiropterans), 5 primates (lorisids), 20 rodents, 9 carnivores, 13 ungulates, 1 tubulidentate (aardvark), 2 pholidotans (pangolins), and 1 hyracoid (tree-hyrax), giving a total of 89 species.

The members of this highly diversified mammalian fauna occupy narrowly specialised ecological niches, permitting colonisation by numerous species which between them utilise a large part of the available food resources. Such complexity can only result from a long period of evolution within that environment, which is now more or less closed to penetration by novel species. An example for this is provided by the domestic rat, *Rattus rattus*, which was introduced by man and has remained restricted to human habitations without colonising the surrounding forest. In Madagascar, on the other hand, this same species has completely invaded the forests, which support a less diversified insular fauna of a less 'competitive' nature. Another significant example is that of palaearctic and intertropical birds which hibernate in equatorial regions (Brosset, 1968). Of 36 species identified at Makokou, not one has been observed in the forest, and all occupy areas cleared by human activity.

Within this natural forest ecosystem, the lorisid primates exploit only a minute fraction of the available dietary components. In fact, the population densities of these five species are considerably lower than those of parallel lemur species in Madagascar, which have more diverse diets.

Various authors who have studied the lorisines and galagines, usually basing their interpretations on observation of the choice of food exhibited in captivity, have pointed out that these animals feed on insects, fruits and occasionally gums. Indeed, in the author's laboratory colony in France all of the five Gabonese species have always fared very well on the same basic diet composed of insects (mealworms, crickets and locusts), fruits (banana, apple, pear, etc.) and a supplement of milk. Yet, under natural conditions, these five species exhibit a variety of dietary regimes distinct and typical for each species! In reality, a species capable of feeding on a fairly large range of food-sources can only make use of certain categories in its natural habitat because of ecological competition with other species. The more competing species there are for a given type of food, the more they become differentially specialised for the collection of certain restricted categories of that food type.

It soon became apparent at Makokou that each of the five lorisid species, according to the dietary categories exploited, occupied a particular spatial section of the forest (particular biotope, canopy zone, undergrowth zone, etc.). This situation, based principally on the height of the forest stratum exploited and on the nature of the supports utilised in locomotion (branches, lianes, etc.), has direct implications for locomotor adaptation and, in certain cases, for systems of defence against any predators which are likely to be encountered. In addition, various behavioural, physiological and morphological specialisations have developed, permitting each species to derive maximum profit in terms of the specific dietary components exploited. These different interactions are illustrated in schematic form in Fig. 16, including the influence of predators which may exert an effect on the evolution of the various behavioural adaptations. This schema, which shows the interaction of biological processes associated with feeding and locomotion, corresponds to the concept of the ecological niche defined by Odum (1953). The niche concept covers the integration of morphological adaptations, physiological responses and specific behaviour patterns contributing to the particular status of an organism within its ecosystem.

In this account of the lorisid primates, attention will be directed primarily to aspects of behaviour directly or indirectly associated with the specific ecological conditions of each of the five species (behavioural ecology). In order to clarify the different aspects as much as possible, successive consideration will be given in Chapters 2 to 5 to diet, locomotion, anti-predator systems, circadian rhythms (rhythms of activity over each 24-hour period), diurnal resting sites, and population parameters. However, since all of these aspects are interrelated, it will be necessary here and there, to assist understanding of certain details, to mention individual features covered in detail in other chapters.

In Chapter 6, covering social behaviour, we shall be faced with behavioural problems concerning relationships between the individuals of each species. In examining social organisation with respect to inter-individual relationships, sexual behaviour, mother-infant relationships and reproduction, information on the behavioural ecology of each species will provide a more reliable basis for interpretation.

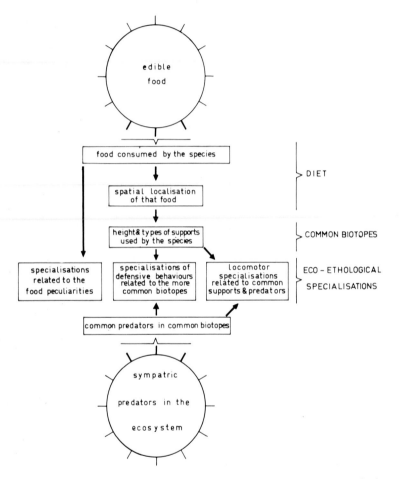

Figure 16. Schematic diagram illustrating the repercussions of the environment (nature of supports and biotopes utilised; available food; predation) on the various behaviour patterns of the lorisids. The adaptive responses of the behavioural repertoire constitute the subject-matter of 'behavioural ecology'.

2

Diet

A. METHODS

Analysis of stomach and intestinal contents has permitted quantitative evaluation of the dietary regime of each species (Table 3). In Gabon the prosimians are hardly ever hunted, and their relative abundance allowed collection of specimens for analysis without menacing the natural populations. Hunting with a rifle proved to be the best method for collecting specimens, since it permitted immediate histological fixation of material accompanied by recording of the exact time of collection. Since digestion was rapidly arrested after collection in this way, information on digestive rhythms could be obtained. by examining alimentary tracts collected at different times during the night. By carefully weighing fresh stomach contents it was possible to evaluate the proportions of different components included in the diets of each species. However, identification of these components can only be conducted at a gross level since the food is subjected to mastication before swallowing. Classification of stomach contents was thus limited to the broad identification of fruits, gums, leaves, buds, wood fibres, fungi and animal prey. With the latter category, analysis of stomach contents permitted more precise determination, though only at the level of the large taxonomic categories of animal prey: beetles (coleopterans), caterpillars and moths (lepidopterans), grasshoppers and locusts (orthopterans), ants and other hymenopterans, isopterans, centipedes and millipedes (myriapods), spiders (arachnids), slugs and snails (gasteropods), and frogs (batrachians). Some fruits can be identified when the kernels or pips are swallowed as well, but identification is often impossible.

The percentage figures calculated (Table 3) are based on overall averages for stomach contents collected regularly throughout the year and at all times during the night. In fact, there are variations according to the season and (sometimes) according to the time of the night, and these may bias the results when such overall averages are used.

In analysing the diet in terms of such broadly defined dietary categories, no differences were found between the diets in primary and secondary forest for any of the species. The proportions of fruits, gums and animal prey remain constant for each species, and this applies to the different categories of animal prey. For this reason, the results for each species in both types of forest have been pooled in Table 3.

The insects and fruits found in the stomach contents were separated with forceps and weighed while still fresh. With the gums, which remain in the stomach only for a short period and then accumulate in the caecum, the total quantity extracted from the stomach and from the caecum was taken into account. The data given in Table 3 represent averages calculated after examination of all dissected specimens.

Table 3 Average fresh weights of the principal dietary components found in stomachs* of the five prosimian species

	Galagines			Lorisines	
	G. demidovii N = 55	*G. alleni* N = 12	*E. elegantulus* N = 52	*P. potto* N = 41	*A. calabarensis* N = 14
Insects †	1.16g	2.22g	1.18g	3.40g	2.00g
Fruits	0.30g	9.20g	0.25g	21.00g	0.30g
Gums	0.15g	negligible	4.80g	7.00g	none

* Gums do not remain in the stomach for very long, however, and the weights in this case are calculated from combined contents of the stomach and the caecum.

† It should be noted that the weight of insects in the stomach contents is approximately the same for all five prosimian species, and that the larger-bodied species complement their diet with fruit and/or gums.

Hence, they do not indicate the daily intake (since the animals eat and digest throughout the night); but they do give an idea of the relative quantities of different dietary components, comparable between species.

B. DIETARY COMPOSITIONS

Fig.17 provides a summary of the dietary preferences of the five prosimian species. At first sight, one can distinguish three which inhabit the canopy – one 'insectivore' (*Galago demidovii*), one 'frugivore' (*Perodicticus potto*), and one 'gummivore' (*Euoticus elegantulus*) – and two which occur in the undergrowth – one 'frugivore' (*Galago alleni*) and one 'insectivore' (*Arctocebus calabarensis*). It will be seen from Chapter 3 that, among the galagines, *Galago demidovii* and *Euoticus elegantulus* are typical inhabitants of the canopy and *Galago alleni* is confined to the undergrowth, while among the lorisines, *Perodicticus potto* occupies the canopy and *Arctocebus calabarensis* occurs in the undergrowth.

In addition to the analysis of stomach contents, direct observations conducted in the forest (Fig.18) permitted frequent collection of data on actual feeding behaviour. This provided a broader basis for the interpretation of the gross categories of material found in the stomachs, and confirmed the distinctions indicated in Fig.17.

Preliminary analysis indicates, then, that different dietary regimes, combined with differential stratification of the species in the forest, largely eliminate competition for food between the five prosimian species. But in fact there is some degree of overlap between the species even at this level of analysis, and finer separation of the principal dietary components (animal prey; gums; fruits) is necessary for better appreciation of the ecological distinctions concerned.

C. ANIMAL PREY

The percentage figures shown in Fig.17 for animal prey ingested by each prosimian species are only relative. However, if one takes into account the absolute quantities of the different dietary components (Table 3), it emerges that each species consumes about the same weight of animal prey. For

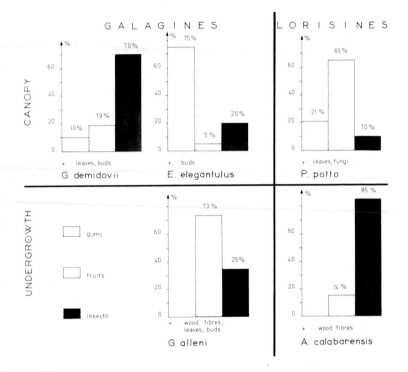

Figure 17. Constitution of the diets of the 5 prosimian species in Gabon. In
order to illustrate the ecological separations of the various species, the
galagines and lorisines have been given two distinct vertical columns,
whilst horizontal columns distinguish animals of the undergrowth from
those of the canopy. The percentage figures given for each dietary
category relate to the total weight of stomach contents (including the
contents of the caecum in the case of gums). In fact, it is necessary to
take account of differences in adult body-size between the prosimian
species. All 5 species consume approximately equal quantities of insects,
and the larger species accordingly complement their diet with fruit
and/or gums (see Table 3).

Figure 18. Two assistants photographed at night during a sortie into primary forest. Note the powerful headlamps adapted from miners' lamps.

example, with *Galago demidovii* (average body-weight = 65 g) 1.16 g of animal prey amounts to 70% of the stomach contents, whereas with *Euoticus elegantulus* (average body-weight = 300 g) 1.18 g amounts to only 20%. The same relationships are found with the lorisines: an average of 2 g of animal prey for *Arctocebus calabarensis* amounts to 85% of the stomach contents, while 3.4 g of animal prey for *Perodicticus potto* represents only 10%.

In fact, for animals which hunt the same prey species using the same technique, the probability of prey capture is the same for a given distance covered, whatever the body-size of the predator. An animal of small body-size like *Galago demidovii* can feed almost exclusively on what it obtains by hunting, but a larger species must necessarily complement its diet with gums (*Euoticus elegantulus*) or with fruits (*Galago alleni*). The same applies to the two lorisines: *Arctocebus calabarensis* (average body-weight = 200 g) is able to feed almost exclusively by hunting animal prey, whereas *Perodicticus potto* (average body-weight = 1000 g) must supplement its insectivorous diet with both fruits and gums.

Experimental tests in the laboratory show that all five prosimian species will feed almost exclusively on insects if provided with a diet of fruit and insects *ad libitum*. This indicates that all species prefer insects if they are available. Support for this is provided by observations in the field. For example, *Galago demidovii* eats fruits predominantly during the early hours of the night when still very hungry; between 18.30 hrs and 24.00 hrs, fruits and gums make up 35% of the ingested food, whilst these components only amount to 20% between 24.00 hrs and 6.00 hrs. A recent study conducted by Beerten-Joly et al. (1974) has demonstrated the presence of chitinase (an enzyme attacking the integument of insects) in the digestive tract of the potto at a level comparable to that in the mole! This shows that despite the large quantity of fruits in its diet the potto is a typical 'insectivore'.

A comparable phenomenon was observed by Hladik and Hladik (1969) with South American monkeys in Barro Colorado (Panama zone). All of the various monkey species ingest the same total weights of insects, which is sufficient for the smallest species but must be complemented with fruits and/or leaves in the diets of the larger species.

Accordingly, there is some suggestion of interspecific

competition for animal prey between the five prosimian species at Makokou. Table 4 shows, in order of importance, the different categories of animal prey found in the stomach contents of the five species. It is at once evident that the two lorisine species feed largely on 'unpalatable' animal prey usually ignored by the galagines: irritant caterpillars, ants of the genus *Crematogaster*, millipedes, etc. Since the galagines and the lorisines are distinguished primarily by their mode of locomotion, it is to be expected that their hunting techniques – and hence the kinds of prey captured – would also exhibit fundamental differences.

(i) *Prey capture by the lorisines*

The potto and the angwantibo are exclusively slow-moving climbers. In the course of their nocturnal excursions, they continuously sniff around them and discover the majority of their prey by their odour. In captivity, all insects are detected by smell, even if they are concealed in a filter-paper container or behind a metal plate with fine perforations. An angwantibo will scarcely respond to insects presented in a sealed plastic container or to silhouettes of insects drawn on paper, but the same silhouette will immediately be located and evoke predatory responses if it is impregnated with the odour of a caterpillar. Angwantibos have been observed hunting in foliage, where they will frequently and unhesitatingly locate caterpillars, which are in most cases hidden beneath a leaf, more than a metre away. The potto has better eyesight than the angwantibo, but its highly developed sense of smell is similarly predominant both in searching for animal prey and in finding fruits. A potto can detect an immobile cricket hidden from view at a distance of one metre.

The technique of prey-capture is also fundamentally similar in the angwantibo and the potto. If the prey animal is small, it is carried to the mouth in the hand and rapidly killed with a few bites delivered to the head and thorax. If the prey is relatively large in size, it is clamped down with one hand and killed by biting before it is taken into the mouth. Various types of capture are utilised, according to the type of prey involved. An angwantibo attempting to capture a moth will rear up on its hind limbs and seize its prey at the bases of the wings with

Table 4 Animal prey of the five prosimian species in Gabon

	Galagines			Lorisines	
	Galago demidovii (N = 55)	*Galago alleni* (N = 12)	*Euoticus elegantulus* (N = 52)	*Perodicticus potto* (N = 41)	*Arctocebus calabarensis* (N = 41)
	small beetles—45%	medium-sized beetles—25%	orthopterans—40%	hymenopterans (ants) —65%	caterpillars—65%
	nocturnal moths—38%	snails—15%	medium-sized beetles—25%	large beetles—10%	beetles—20%
	caterpillars—10%	nocturnal moths—15%	caterpillars—20%	snails—10%	orthopterans
	hemipterans	frogs—8%	nocturnal moths—12%	caterpillars—10%	a few dipterans
	orthopterans	ants—8%	a few ants	orthopterans	a few ants
	centipedes	spiders—8%	a few bugs (homopterans)	millipedes	—
	a few bugs (homopterans)	orthopterans	birds*	spiders	—
	a few pupae	termites	—	termites	—
	—	centipedes	—	birds*	—
	—	pupae	—	bats*	—
	— is	a few caterpillars	—	—	—

Note:
The percentage figures are relative to the total mass of food identified as 'animal' in the stomach contents (fresh weight). For each prosimian species, the prey categories are ranked in order of importance from top to bottom. Prey types marked with an asterisk (*) have been included not on the basis of stomach content analysis but following direct observations conducted in the forest. (Small vertebrates captured from time to time represent only a minor component of the natural diets.)

both hands. Caterpillars are flattened against the support or simply detached from a leaf with one hand. The angwantibo subsequently grasps the caterpillar's head in its mouth and wipes its hands along the prey's body for 10-20 seconds before eating it. This vigorous 'massage', which removes most of the hairs from the caterpillar's body, seems to be a special behavioural adaptation for the angwantibo's diet, which includes almost all caterpillar species, many of which are ignored by the other prosimians. Whenever caterpillars with irritating properties are captured, they are massaged extensively and, after they have been ingested, the hands and the snout are wiped along branches for some time, often for more than a minute. The same massaging behaviour is elicited by other categories of prey, including hairless caterpillars and orthopterans (locusts, grasshoppers, etc.). An innate behaviour pattern is apparently involved, since massaging appears in captivity even before the angwantibo is capable of capturing insects, as soon as proffered prey are eaten at an age of 2 months.[1]

The potto also captures slow-moving, large prey by flattening them against the support (large beetles, millipedes, etc.). Ants, which constitute the essential component of the potto's animal food, are simply lapped up with the tongue along the branches where moving columns are present. Some of the pottos dissected had stomachs packed with ants of the genus *Crematogaster*. The potto also eats prey with repulsive odours such as millipedes (*Spirostreptus sp.*) or pungent-smelling orthopterans (*Zonoceus variegatus*), and occasionally it will attack and kill small vertebrates, such as a young bat (*Epomops franqueti*) which was observed in the hands of a potto on one occasion. On two occasions, a female potto was observed in the study area visiting a colony of weaver-birds (*Ploceus cuculatus*). These birds construct a large number of nests, each of which has an opening situated at the end of a downward-pointing neck. In one case, the potto was actually seen suspended from its hindlegs near to one of these nests and was doubtless in the process of exploring its contents. On another occasion, some villagers brought along a potto which had been surprised eating a chicken in their chicken-coop!

[1] Similar behaviour is exhibited by the toad and the mole before ingestion of earthworms.

Small vertebrate prey are seized with one hand, which is moved extremely rapidly and always grasps the prey by its back. In captivity, birds presented to pottos are eaten together with their feathers whereas with mammal prey at least a part of the skin is left uneaten (Walker, 1969). Walker interprets the slow-moving locomotion of the lorisines as an adaptation linked, among other things, to the capture of birds. The potto readily eats birds when these are presented in captivity, but under natural conditions capture of such prey must be relatively rare. Study of the stomach contents reveals that slow-moving, easily captured arthropod prey constitute the bulk of the animal diet of the lorisines, at least in the area where the present field-study was carried out. It will be seen in Chapter 3 (Locomotion and Defensive Behaviour) that the slow-moving locomotion of the lorisines themselves is particularly adapted for concealment from predators.

Thus, under natural conditions, the potto and the angwantibo consume considerable quantities of irritant or pungent-smelling prey. Yet this specialisation in diet is not due to preference, since in captivity these two prosimian species will leave untouched ants and irritant caterpillars when presented at the same time with sphingid moths and orthopterans. Accordingly it would seem that the lorisines are behaviourally specialised for hunting certain categories of prey ignored by the galagines, concentrating on slow-moving, easily captured prey which are protected against many predators by their irritating hairs or by repugnant glandular secretions, and which can generally be perceived from some distance away because of their strong smell. It seems probable that this dietary adaptation has been developed in parallel with the locomotor specialisation of these prosimians, which are unable to cover large areas in search of insects. The same peculiarity is exhibited by a third lorisine species (*Loris tardigradus*) in Ceylon (Petter and Hladik, 1970).

(ii) *Prey capture by the galagines*

The bushbabies move around very rapidly by a mixture of running and leaping. The resulting commotion in the foliage often provokes insects to flee, and they are then exposed to auditory or visual localisation. In captivity, a cricket can escape detection in litter on a cage-floor as long as it remains immobile (e.g. when anaesthetised by cooling), but a bushbaby will spot it as soon as there is the slightest movement. When a moving prey is present, bushbabies direct their mobile ears in the appropriate direction, and observations have shown that they can localise perfectly sounds emitted by insects. Flying locusts and scurrying crickets can be localised on the other side of a plywood screen, and bushbabies will follow the movements of these insects with their heads, just as if they could actually see the prey. A simple experiment can be carried out by suspending a locust from a thread and moving it behind an opaque plywood screen. When the insect flutters its wings, a bushbaby follows its movements by orienting its ears in the appropriate direction, and follows the insect by running along behind the screen. This experiment has been repeated 20 times with both *Galago demidovii* and *Euoticus elegantulus*, with consistently conclusive results.

Auditory prey-detection of this kind is very important, for bushbabies very often catch insects as they take off. In order to do this, a bushbaby will propel its body forwards whilst maintaining a grasp on the branch with its hind feet. After the prey has been captured, the bushbaby then returns to the squatting position in a manner which varies from species to species. *Galago demidovii* recoils immediately to its starting position, even when it has previously extended its body horizontally. On the other hand, *Euoticus elegantulus* (a much heavier animal) is unable to return immediately to its initial posture. After capturing its prey, the needle-clawed bushbaby ends up suspended head downwards by its hindlegs and it only returns to the squatting position on the branch after transferring the prey to its mouth. This behaviour has never yet been observed with *Galago alleni*, however. Allen's bushbaby hunts primarily at ground level, and it is primarily the two species inhabiting the canopy which are suited to capturing insects in flight. Lepidopterans and orthopterans, which are very apt to take off when startled, are a major constituent of their animal prey. It should be noted that *Microcebus murinus* (the lesser mouse lemur of Madagascar) also captures insects in flight, using the same technique as *Galago demidovii* (Martin, 1972a). Bushbabies also capture insects sighted on branches used as supports during locomotion or on nearby branches. In such cases, they approach the prey slowly and then, having brought the body into a crouching position, they pounce rapidly whilst maintaining a hold on the branch with their hind feet. The prey is always captured in the hands first and then transferred to the mouth before the previous body posture is resumed. Insects taken in this way are rapidly killed with bites delivered to the head and thorax, as was found with the lorisines.

On one occasion, a needle-clawed bushbaby was seen eating a small warbler (*Camaroptera brevicaudata*), commencing with the dorsal surface. In captivity, Allen's bushbaby readily eats small vertebrates (frogs, lizards, young rodents, etc.), and a small frog was found in the stomach contents of one individual.

The galagines and lorisines thus discover and capture their prey in different ways, resulting in the consumption of quite different animal types. The lorisines take slow-moving and relatively unpalatable prey, whilst the galagines take more rapid, generally palatable prey. The two groups of prosimians do not therefore enter into competition in their search for insect food, and any significant competition would be expected between the three bushbaby species on the one hand and the two lorisine species on the other. In fact, within the two subfamilies (Galaginae and Lorisinae) stratification of the ecological niches within the forest already separates the species. *Euoticus elegantulus* hunts in the sparse foliage of the canopy layer, *Galago demidovii* hunts in the dense foliage associated with patches of invading lianes, and *Galago alleni* hunts on the forest floor. Of the two lorisines, *Perodicticus potto* hunts its prey in the canopy, while *Arctocebus calabarensis* hunts among the small trees of the undergrowth (see Chapter 3). In addition to such separation through stratification, Table 3 shows for each prosimian species a predilection for certain prey types: in the canopy, *Euoticus elegantulus* hunts orthopterans, *Galago demidovii* hunts small beetles and nocturnal moths, and *Perodicticus potto* concentrates heavily on ants of the genus *Crematogaster*; whilst in the forest understorey *Galago alleni* hunts prey of various kinds on the ground and *Arctocebus calabarensis* captures caterpillars, which are in most cases of an irritant type.

D. GUMS

Gums represent a dietary category which is relatively little utilised by vertebrates. To the author's knowledge, apart from one South American monkey, the pygmy marmoset (Dr. M. Moynihan, pers. comm.), and a didelphid marsupial of the genus *Philander* (G. Dubost, pers. comm.), only the lemurs and lorises seem to be specially adapted for this food source. Gums are actually eaten by *Galago senegalensis* (Sauer and Sauer, 1963), *Galago crassicaudatus* (Bearder and Doyle, 1974), *Euoticus elegantulus*, *Galago demidovii*, *Perodicticus potto* (Charles-Dominique, 1971b), *Microcebus murinus* (Martin, 1972a), *Phaner furcifer*, and *Microcebus coquereli* (Petter et al., 1971). However, it is only in *Euoticus elegantulus* in Africa and *Phaner furcifer* in Madagascar that gum provides the principal food-source. Martin (1972b) interprets the original evolution of the characteristic 'tooth-scraper' of the Strepsirhini (lemurs + lorises) as an adaptation linked to collection of gums, which often necessitates such a 'precision tool' for extraction from fissures in trunk surfaces. According to this view, the well-known grooming role of the tooth-scraper (Andrew, 1964; Buettner-Janusch & Andrew, 1962) in lemurs and lorises would represent a secondary function.

Gums vary greatly in their chemical composition, and only a few of the available kinds are actually eaten by the prosimians at Makokou (exudations of *Entada gigas*, *Entada celerata*, *Pentacletra eetveldeana*, *Piptadenastrum africanum*, *Albizia gummifera*; all five species – two lianes and three trees – belonging to the Family Mimosaceae). All other available gums are ignored.

Investigations of the chemical composition and digestion of gums, which are still in progress (Hladik and Charles-Dominique, unpublished data), indicate that the gums concerned consist of polymerised chains composed of galactose and rhamnose. The gum exudates appear in the form of fine droplets of a gelatinous consistency which gradually increase in size. They are most commonly found at the bases of fissures in the bark or at places (usually on the trunk) where the tree has suffered damage at some time in the past. Homopteran bugs are responsible for a great deal of gum-exudation, which represents the plant's natural defensive response against the bite-wounds inflicted by the bugs' mouthparts. As it happens, a detailed study carried out by Boulard (unpublished data) on prosimian stomachs containing gums has revealed the presence of various insects which were apparently ingested along with the gums: *Oxyrachis sp.* (Membracidae), *Panka sp.* (Tibicinidae) and *Mnemosyne sp.* (Tropiduchidae) in stomachs of *Euoticus elegantulus,* and *Epilemna sp.* (Ricaniidae) in the stomach of *Galago demidovii.*

Although all three prosimians in the canopy eat gums (Fig.17), only the needle-clawed bushbaby and the potto can conceivably be in competition with one another, since Demidoff's bushbaby only eats very small quantities of gums from time to time (see Table 3). We therefore need to examine *Euoticus elegantulus* and *Perodicticus potto* in turn to see how they collect gums and to determine whether there is any competition between these two species.

(i) *Euoticus elegantulus*

Throughout the night, the entire activity of the needle-clawed bushbaby is directed primarily towards the search for gums. The individuals of a group, which always sleep together in the same area during the daytime, separate at dusk and each bushbaby sets off in a defined direction. Subsequently, each individual visits lianes, branches and trunks which produce gums, collecting tiny droplets at each site. For each individual animal, the itinerary covered is the same from night to night, permitting familiarity with the location of all points of gum-exudation. The author has actually been able to observe individual *Euoticus* in the process of visiting certain wound areas on trees where tiny droplets of gum were forming, and collecting the exudate by licking or scooping with the tooth-scraper. These sites of gum production were visited nightly, so that the exudates were always collected at the very beginning of droplet-formation.

When collecting gum from lianes of the genus *Entada, Euoticus elegantulus* runs rapidly along the long, serpentine stems at the upper level of the understorey just prior to their entry into the canopy. With such lianes, the gum production points are not stable, but the bushbaby localises them by smell as soon as they come within a range of 20-30 cm. Thus, one can often see needle-clawed bushbabies running along a liane stem and abruptly stopping to collect a gum droplet, which may lie out of the line of vision on the other side of the stem. When following *Euoticus elegantulus* in the forest, the author counted an average of 100 gum-collections an hour. Making allowance for brief resting-periods during the bushbaby's nocturnal activity phase, this indicates collection of gum at 500-1000 sites per night. The plant producers of gum are visited one after another in sequence at regular times as well as along a standardised itinerary. For example, at Ipassa one female needle-clawed bushbaby observed over a week arrived each night at between 20.40 hrs and 21.00 hrs to visit a liane (*Entada celerata*), passed directly from this food-source to a tree (*Albizia gummifera*) and then moved on to another liane (*Entada gigas*) at about 22.00 hrs. She thus took about $1\frac{1}{2}$ hours to cover a distance of 200 metres. This regularity is due to the fact that individuals of *Euoticus elegantulus* are very well acquainted with the positions of gum-producing plants which exude gum throughout the year. This situation precludes any search for new sources of food of the kind found with frugivorous species, whose food resources fluctuate according to the season.

In addition to feeding behaviour which is finely adapted to a diet of gums, *Euoticus elegantulus* also exhibits a number of anatomical peculiarities correlated with this particular food-source. First of all, the tongue and tooth-scraper are relatively longer and narrower than in the other four sympatric prosimian species (Fig.19), and this is undoubtedly of advantage in reaching relatively inaccessible gum exudations in crevices. Another peculiarity, which gives this bushbaby its common name (needle-clawed bushbaby), is that each of the nails bears a keel, terminated by a short point 1 mm in length, which enables this species to climb on large trunks and branches inaccessible to the other prosimians. *Euoticus elegantulus* have frequently been observed, whilst searching for gums, moving head downwards on broad tree-trunks with a diameter in excess of 70 cm (see the discussion of locomotion in Chapter 3).

(ii) *Perodicticus potto*

The potto eats gums in a much less regular fashion than the needle-clawed bushbaby, and its diet consists largely of fruits. Whereas all of the *Euoticus* dissected (N = 62) contained gums

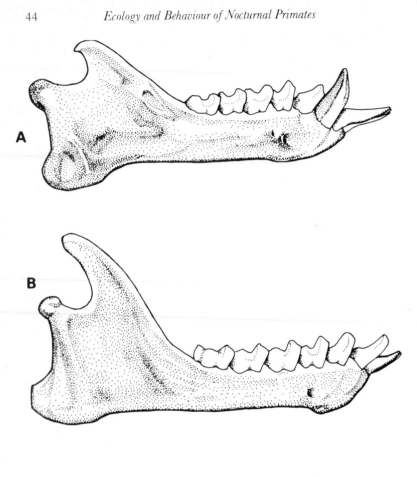

Figure 19. Lower jaws of *Euoticus elegantulus* (A) and *Galago alleni* (B) compared to show the great development of the tooth-scraper ('comb') of *Euoticus*, as an adaptation for gum-eating.

in the digestive tract, 30% of the pottos examined had none at all, and only 50% had eaten appreciable quantities (N = 41).

The gums found in the stomachs of pottos occurred as large, compact aggregations, whereas stomachs of *Euoticus elegantulus* contained fine droplets aggregated in clumps. The slow locomotion of the potto prohibits the numerous collecting visits seen with the needle-clawed bushbaby, and the potto is thus constrained to eat large lumps of gums which are already relatively stale and are usually ignored by *Euoticus elegantulus* for this reason.

It is hence apparent that these two prosimian species collect gums with quite different techniques, and it is at once clear that the needle-clawed bushbaby is best fitted, in terms of both its anatomy and its behaviour, in any competition which may arise. However, the system of collection would seem to be uneconomical from the point of view of the relationship between energy expenditure and energy gain. More widely spaced visits would permit the accumulation of gums at the points of production, thus reducing the required distance to be covered for collection of a given quantity. The lack of economy would seem to be the result of a long period of inter-specific competition: the more a species visits points of production during any one night, the less chance there is for competing species to discover the accumulated gums in sufficient quantity. A parallel advantage of this system of regular collection is that an animal which frequently visits a large number of gum-production points will learn their location with great precision. The regular itineraries followed permit achievement of maximum efficiency in locomotion, without the loss of time and energy that would be occasioned by a search 'at random'.

E. FRUITS

Determination of fruit species from the kernels and seeds ingested and included in the stomach contents is often very difficult. For this reason, it is only possible to give the names of a few species eaten by the prosimians at Makokou. In secondary forest areas, the fruits of the parasol tree (*Musanga cecropioides*) are consumed throughout the year by the prosimians as well as by numerous other mammals and birds. Trees of the genus *Uapaca* are abundant in the flooded forest areas and produce fruit primarily in the period October–April. *Perodicticus potto*, *Galago alleni* and *Galago demidovii* all eat these fruits quite frequently. In addition, seeds of *Ricinodendron africanus* have been found in potto stomachs, while *Galago alleni* has been directly observed eating fruits of *Polyaltia suaveolens* (Annonaceae), *Diospyros piscatoria* and *Diospyros hoyleana* (Ebenaceae). As a general rule, the fruits eaten by the prosimians are soft and sweet. However, the potto is able to deal with fruits surrounded by a hard shell, though only the soft part contained inside is actually eaten. *Galago demidovii* occasionally profits from this and may arrive to eat the rest of the pulp after a potto has passed by and opened a hard-shelled fruit.

Perodicticus potto and *Galago alleni* are the main consumers of fruit, but their ecological separation eliminates any competition for food between them. The first species eats fruit exclusively in the trees, whilst the latter feeds on fruit which has fallen to the ground and in small fruiting bushes in the undergrowth. In an enclosure containing small bushes, we presented these two prosimian species with a choice between eating bananas placed on the ground and bananas attached to branches. *Perodicticus potto* always selected bananas attached to branches, whilst *Galago alleni* always chose the fruit placed on the ground.

Allen's bushbaby eats fruit in a particular fashion, since it succeeds in swallowing fruits up to the size of a cherry without mastication. It simply raises its head to aid swallowing, and the fruit slips down untouched. As a result, one often finds entire *Uapaca* seeds in the stomach contents of this species. In this way, *Galago alleni* can rapidly fill its stomach in a very short space of time, reducing exposure to predators to a minimum. It is generally true of frugivores that they do not stay for very long in the area where they collect their fruits. Fruits are collected rapidly and then eaten or digested in a sheltered spot (cf. the cheek-pouches of Old World cercopithecine monkeys and of some rodents, the crop in birds, and the rumen in the stomach of some ruminants). The behaviour of *Galago alleni* seems to fit the general picture. This species can swallow a large quantity of fruits, and its stomach can contain up to 20 g, amounting to about 8% of its body weight, whilst the other bushbaby species never show stomach fruit contents amounting to more than 3% of their body weight in any one night. *Euoticus elegantulus* and *Galago demidovii* eat fruit slowly, picking off tiny pieces with their tooth scrapers and never swallowing the kernels like *Galago alleni*. Since the needle-clawed bushbaby and Demidoff's bushbaby feed only in the trees, they are apparently exposed to less risk while eating than is the case on the ground.

The potto also exploits fruits in the canopy, but its slow-moving locomotion increases the risks of exposure to predators. Its slow progression constitutes an excellent means of concealment under a cover of vegetation, but fruits are often located at the tips of branches, where foliage is sparse. The potto can swallow entire fruits (*Uapaca sp.*) or very large chunks of fruit, and remains in fruit-trees for a minimum period of time. A banana can be eaten completely in less than half a minute, and the highly extensible stomach sometimes contains up to 75 g of fruit – that is to say 8% of its body weight, just as in Allen's bushbaby. It is rare for a potto to be found in a tree in fruit; usually, animals of this species are found 20 metres or so away in a leafy tree with their stomachs full of fruits.

Detection and exploitation of fruit

In the forest, fruits are distributed according to a diversified and fluctuating pattern. Large trees in fruit are the easiest to locate: They attract a great number of birds and monkeys, and fruit which falls to the ground often allows sighting of such a tree from some distance away. In many cases the enormous quantity of fruit produced cannot be completely exploited by the animals which are attracted. However, in Gabon massive production of fruit of this kind is relatively rare; the trees concerned are widely dispersed in the forest and fruit-production does not usually last more than a week. It is more generally true that fruits are found in smaller localised quantities than this, but that the trees bearing them are more numerous in occurrence. Trees producing fruit in this way are usually exploited in a regular fashion, though they often pass unnoticed. Despite this drawback, they play an essential part in the feeding of nocturnal mammals of small body-size.

In order to obtain some estimation of the proportions of fruits produced in small batches in relation to those produced in large batches, counts were carried out along a standard pathway 3 km long, covered 10 times in the course of a year. Fruits are difficult to spot among foliage, and account was taken only of fruit which had fallen to the ground, so no quantitative analysis was possible. However, it emerged quite clearly that trees producing small batches of fruit are the most common (Fig.20).

During the daytime, large fruiting trees are exploited by numerous birds (green pigeons, hornbills, touracos) and by

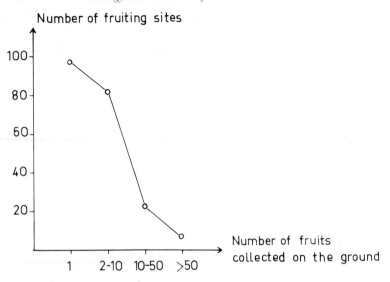

Figure 20. Distribution of fruits in primary forest according to the abundance of the fruiting sites. The counts of fruit are based on fallen fruit. It can be seen that sites with small quantities of available fruit are encountered most often.

cercopithecine monkeys (*Cercocebus albigena*; *Cercopithecus spp.*; *Miopithecus talapoin*). All of these animals are gregarious and have relatively extensive home ranges which permit them to exploit in large groups a fair number of large, dispersed fruit trees which can be spotted from some distance away. At night, on the other hand, one only finds a small number of frugivorous mammals in such trees – with the exception of fruit-bats. Using traps placed close to such trees in fruit, the maximum number of individuals of any given prosimian or rodent species that could be caught was only 2-3. Long-term, regular trapping showed that each individual of the five prosimian species covered a home range which remained relatively stable throughout the year and even from one year to another (see the discussion of social life in Chapter 6). Since these small-bodied, nocturnal mammals have relatively small home ranges, they feed almost exclusively on fruits produced in small quantities and it is only on rare occasions that they come to feed in a large tree in fruit, if it happens to lie within the habitual home range area.

Fruits are discovered by the prosimians in the course of

virtually continuous exploration of their home ranges. A basket of bananas sited experimentally in the forest is soon discovered by prosimians living in the area: Demidoff's bushbaby and Allen's bushbaby take 3-5 days to find the basket, whilst the potto will find it in 5-10 days. Trees in fruit are exploited as soon as they are located by the prosimians, and daily visits are made thereafter, with the animals going directly to the trees concerned at the beginning of the night. Yet a discovery of this kind does not arrest exploration of the home range, and marked individuals have been seen eating successively at different places during the night. Dissection of prosimians taken within the forest (*Perodicticus potto* and *Galago alleni*, the two most frugivorous species) has shown that an individual will often eat fruits of two or three different species in any one night.

Thus, a potto or an Allen's bushbaby will always succeed in obtaining food, even if one of the fruit-bearing trees which is habitually visited should reach the end of its phase of fruit-production or be visited by another animal. Exploration of the home range on a continuous basis ensures, in addition, that trees which are just beginning to bear fruit will be located to take the place of those nearing the end of fruit-production.

F. DIETARY CONDITIONING

From the second month of age onwards, following weaning, young prosimians of all five species accompany their mothers increasingly during the course of the night. Young pottos and angwantibos follow their mothers by walking behind them or by riding on the mother's back; young bushbabies follow their mothers or some other adult in the social group (other females or a dominant male). Initially, the prey-capture responses are not entirely developed and these young animals would be incapable of finding and selecting their food unaided. They grasp fragments of insects from the hands or teeth of their mother with impunity, though she will not release the prey (Fig.21). In most cases, the young animal will tear off a limb of the insect, which will be extensively masticated before being swallowed (observations conducted in captivity on pottos, angwantibos, Allen's bushbabies and Demidoff's bushbabies). The first fruits to be eaten are similarly taken from the mother's mouth. In the potto, an older infant will take fruits directly from

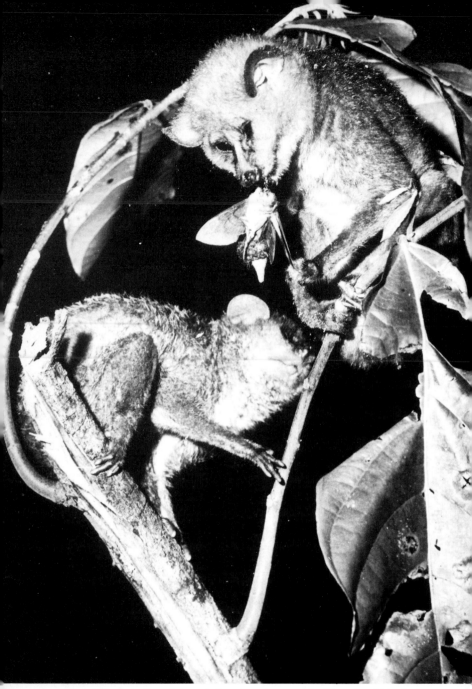

Figure 21. Arctocebus calabarensis female eating a sphingid moth which she has been given. Her infant is approaching to tear off fragments. (Photo: A.R. Devez)

a tree whilst remaining attached to the mother with one or both of its hindlimbs. Under natural conditions, the young needle-clawed bushbaby follows its mother and eats gum whenever she stops to do so. It is doubtless in the course of such learning in the company of the mother that these prosimians become conditioned under natural conditions to feeding on numerous insect species, fruits and gums which will later be incorporated in the overall diet.

The author has hand-reared young pottos and bushbabies, which become tame very rapidly and then re-direct towards the foster-parent behaviour patterns which they normally perform in relation to the mother. In this way, a young potto learned to recognise all of the various kinds of food eaten by the author. Clinging to the shirt, the potto initially tried to intercept food on the way from the plate to the mouth, and it was not long before the animal was eating directly from the plate with its hind feet maintaining a firm hold of the author's shirt-front. As soon as the author moved away from the table, the potto would abandon the plate and return to a full grasp on the shirt. Once this particular individual was weaned, it would accept all left-overs from the kitchen, which were consistently refused by pottos captured as adults.

As a general rule, young prosimians will become habituated rapidly and easily to a new dietary regime, whereas adults (particularly those of advanced age) are much more difficult to rear. Eventually, about thirty captive prosimians of the five species were taken back to France from Gabon. In captivity in Gabon, these animals were fed largely with various insects, including locusts and crickets, captured with the aid of an ultra-violet lamp. In the laboratory in France, however, we were only able to provide mealworms and beetles (*Tenebrio*), and unfamiliar species of locusts and crickets (*Locusta, Gryllus domesticus* and *Gryllus bimaculatus*). These novel insects were refused almost without exception for the first few days, despite the fact that the prosimians had not eaten insect food for several days before arrival, and the first individuals to accept the new food were predominantly young animals. Some individuals took two months or more before they would accept the locusts. The same observation was made with the transition to European fruits (cherries, apples, strawberries, gooseberry jam, etc.).

Conditioning to particular food-types can thus continue after weaning and even persist to the adult stage, but this latter period of learning is more difficult and longer than the stage of a few months spent in proximity to the mother. Of course, this earlier period of conditioning of the infant whilst with its mother (duration: 3-6 months) is indispensable for effective learning of palatable food-types. Numerous insects have glands producing repellent secretions, poison-glands, or irritant hairs, and the author has on several occasions observed accidents which have occurred with young animals reared artificially without their mothers. For example, such young prosimians may capture 'repugnant' moths, though these are released at once and a single experience is sufficient to prevent recurrence. A young Allen's bushbaby which had been hand-reared once captured an irritant moth (Family Thometopoidae) and its snout was swollen for two days thereafter to such an extent that the animal could not take solid food. Under similar conditions a young, hand-reared Demidoff's bushbaby died almost immediately after capturing an irritant moth, even though the prey was rejected a few seconds after contact with the mouth.

Very young prosimians take some time to recognise prey as such, inclining their heads from side to side before capture, on occasions for more than 20 seconds. Insects are always sniffed first and a certain amount of individual conditioning is necessary for young animals, which do not kill their prey as rapidly as adults do. In many cases, insects are allowed to escape, or they are attacked by the abdomen with a consequent risk that the young prosimian's snout may be bitten or scratched. However, after capturing 5-10 insects, prosimians learn to start by attacking the prey's head or thorax.

3

Locomotion and Defensive Behaviour

Equatorial rain-forest, which is comparable to a gigantic heterogeneous meshwork, presents an entire range of supports to arboreal animals and a diversity of pathways for passage from one point to another. Just as there are behavioural and anatomical specialisations linked to diet, one may expect to find other specialisations associated with the particular supports most frequently utilised.

A. HEIGHT AND NATURE OF SUPPORTS

Since the principal activity of prosimians consists of the search for food, their spatial localisation within the forest is obviously dependent upon the distribution of appropriate food-sources, though the actual supports selected for locomotion may be to some extent determined by species preference. The first task, then, prior to the study of the mode of locomotion of each prosimian species, was detailed recording of the supports utilised in locomotion. From the outset, systematic notes were made of the characteristics of supports on which prosimians were observed during the night, using the following categories: type of support (ground, thin trunk, large liane base, large trunk, large branch, foliage, foliage mixed with lianes), height above the ground[1], diameter of the support, orientation of the support (horizontal, oblique or vertical). The technique most frequently used was that of conducting searches and observation with a battery-operated headlamp (Fig.18), amounting to about 2500 hours of night-time

[1] Heights were judged by eye in most cases, but from time to time exact measurements were made with an altimeter in order to maintain accuracy.

observation in the forest during the total of 42 months spent in Gabon. Since the light-beam often had an effect on the subsequent movements of any prosimian observed in this way, account was only taken of the characteristics of the support on which any animal was *first* sighted at any one time.

Figs. 22, 23, 24 and 25 show the data obtained from 836 sightings of prosimians in primary and secondary forest. Later observations were conducted without systematically noting the physical characteristics of supports, but by then it had proved possible to obtain more precise information in other ways about the distribution of certain species. In fact, all direct observation was conducted from ground-level, such that the probability of sighting decreased with the height in the canopy at which any species was moving. For this reason, the collected data indicating the heights of animals in the canopy (Fig.23) distorted the actual state of affairs. The figures must therefore be supplemented with a description, for each prosimian species, of the biotopes and pathways most frequently utilised, incorporating not only information from direct observation but also any indications provided by noting of vocalisations and (particularly) by radio-tracking. (An animal carrying a miniaturised radio-transmitter can be localised at any time, with great precision, by means of a directional antenna. Following the description of the individual species, the typical pathways followed by the five species in the forest will be illustrated with a schematic diagram.

(i) *The potto*

In primary forest, *Perodicticus potto* is usually sighted at heights of 5-30 metres above ground-level. Large branches and lianes provide the preferred pathways. Being an exclusive climber, the potto is obliged to descend large branches to their forks and to climb up finer and finer branches in order to reach to foliage of a neighbouring tree, where the process is reversed. The supports most commonly utilised have diameters of 1-15

Figure 22 (right). Chart of relative frequencies of utilisation of different supports by the 5 prosimian species in primary forest and in secondary forest. The columns indicate percentages of the total counts obtained in the forest. For each class of support, the number of observations concerned is indicated above the relevant column. (N = total number of observations on which percentage figures are based.)

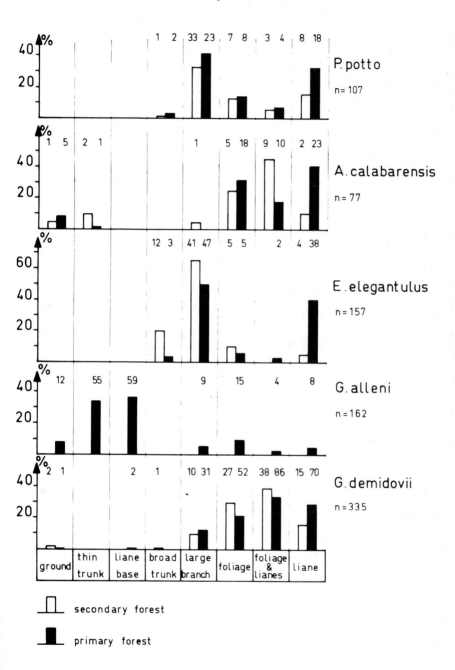

ground | thin trunk | liane base | broad trunk | large branch | foliage | foliage & lianes | liane

☐ secondary forest

■ primary forest

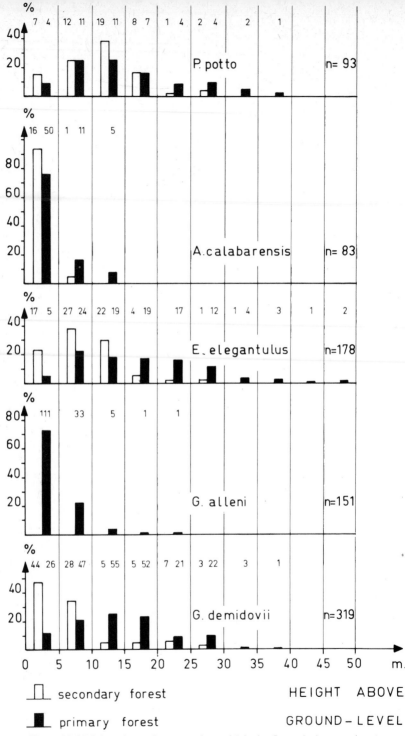

Figure 23. Heights above the ground at which the 5 prosimian species were encountered in primary forest and in secondary forest. (N = total number of observations on which percentage figures are based.)

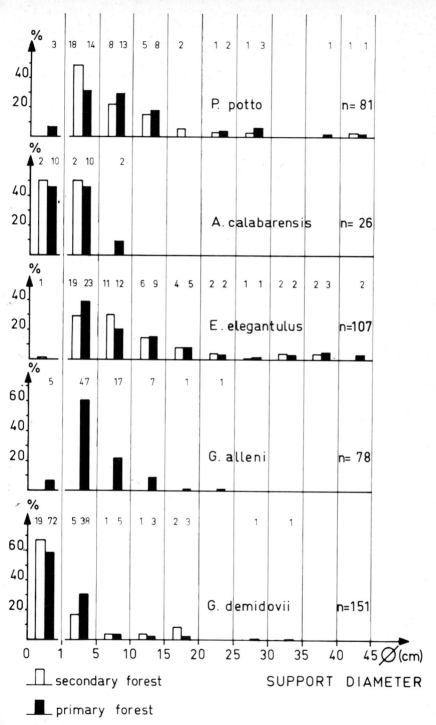

Figure 24. Diameters of the supports used by the 5 prosimian species in primary forest and in secondary forest. (N = total number of observations on which percentage figures are based.)

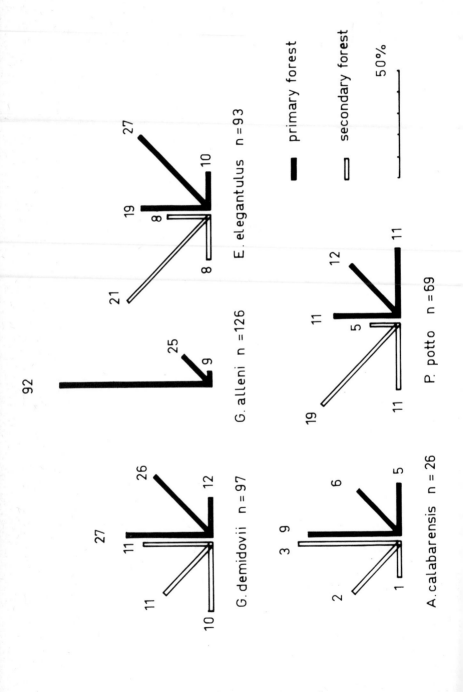

E. elegantulus n = 93

G. alleni n = 126

P. potto n = 69

G. demidovii n = 97

A. calabarensis n = 26

primary forest

secondary forest

50%

cm. In many cases, a liane may provide a short-cut which the potto will normally exploit.

When following two pottos fitted with radio-transmitters in primary forest, it was observed that they spent most of their time in the upper canopy at heights of 20-40 metres. In the course of their progression they sometimes descended considerably, so that they could be seen by the observer, but most of the time they were invisible from the ground. Figure 23 thus underestimates the typical height at which the potto moves in the canopy, as would have been expected in view of the fact that this species is specially adapted for concealment and relies upon discreet locomotion.

In secondary forest, the potto is observed lower down in the trees because of the overall reduction in tree size, but again its locomotion is largely restricted to the canopy. Pottos may occasionally descend almost to ground-level where certain fruits are ripening, but locomotion is difficult in the bushy vegetation at this low level. This slow-climbing species, being essentially frugivorous, typically searches for fruits in zones where the sun permits flowering and fruiting of trees and lianes, that is to say predominantly in the upper regions of the canopy.

(ii) *The angwantibo*

Arctocebus calabarensis was never spotted at a height of more than 15 metres in primary forest, and it was most often observed at a height of 0-5 metres in the undergrowth. This would seem to be characteristic of the species, for Jewell and Oates (1969) observed the same behaviour with a different subspecies (*A. c. calabarensis*) in Biafra. The angwantibo feeds almost exclusively on caterpillars and explores the foliage of bushes in the undergrowth by using small lianes between the bushes to pass from one to another (Figure 26). The typical diameter of the supports used, lying in the range of 0-5 cm, is

Figure 25 (left). Orientations of supports used by the 5 prosimian species in primary forest and in secondary forest. The orientations of the columns represent the orientations of the supports: *vertical* (75°-90°), *oblique* (15°-75°), and *horizontal* (0°-15°) with respect to the horizontal plane. The length of each column is proportional to the percentage utilisation of supports of that orientation. (N = total number of observations on which percentage figures are based.)

A

Figure 26 A and B. The angwantibo, *Arctocebus calabarensis*, in its natural
   habitat: undergrowth of primary forest in a zone of dense vegetation
   produced following tree-falling. (Photo: G. Dubost)

in accord with the angwantibo's small body-size. Small lianes
in the undergrowth are much more abundant in old clearings
where the penetration of light encourages their growth
following the fall of a large tree. It will be seen in Chapter 5
that in primary forest *Arctocebus* is closely associated with
forest zones where tree-falling is common ('tree-fall zones'
exposed to tornadoes because of their particular topographical
situation). Angwantibos were almost always encountered in
these areas at Makokou, and they almost always managed to
flee from the small trees by using the fine lianes entangled
with them. Occasionally, this prosimian species was surprised
on the ground and in fact in the course of the study five
individuals were obtained which had been killed in traps set
on the ground on tracks habitually used by duiker (forest
antelopes) and other terrestrial mammals in primary forest.

B

(iii) *Demidoff's bushbaby*

This small-bodied bushbaby lives in very dense vegetation where other prosimians would have great difficulties in moving. The supports used by this species are usually less than 1 cm in diameter and are highly variable in orientation. The preferred biotopes are zones with fine branches and areas of foliage invaded by lianes (liane curtains), in both primary and secondary forest. Such curtains of lianes hang down for a distance of several metres below the area of invaded branches and the subsidiary stems of the lianes are intertwined, forming

Figure 27. Demidoff's bushbaby, *Galago demidovii*, among fine lianes. (Photo: C.M. Hladik)

a dense lacework on top of which the foliage regenerates. These liane aggregations grow in size from year to year, forming a mass 1-2 metres thick and 3-5 metres in height. Liane curtains formed in this way are common low down in young secondary forest, along the banks of rivers and in 'tree-fall zones'. In primary forest, such curtains are usually found high up in the canopy (Fig.27).

In secondary forest and in tree-fall zones of primary forest, Demidoff's bushbaby moves around in bushy vegetation a few metres above ground-level, sometimes descending quite close to the ground. This prosimian species is frequently encountered at roadsides and in dense vegetation in ditches (*Aphranomum sp.* and small lianes), with the result that several authors (e.g. Cansdale, 1960; Jones, 1969) concluded that it usually occurs low down in the forest. However, Vincent (1969) reports that *Galago demidovii* lives high up in the forest block of the Congo. In primary forest, where they are abundant, Demidoff's bushbabies move around at between 5 and 40 metres above ground-level, passing through dense foliage and liane curtains. But, without powerful illumination and/or a good knowledge of the characteristic vocalisations of this species, it is easy to overlook them at that height.

(iv) *Allen's bushbaby*

As with the angwantibo, this species is found in the undergrowth of primary forest, but its distribution is independent of the 'tree-fall zones'. In the majority of cases, *Galago alleni* was encountered at heights of 1-2 metres on vertical supports 1-15 cm in diameter provided by slim tree-trunks or the vertical bases of large lianes (Fig.28).

In primary forest, some trees are completely invaded by lianes, whose stems are attached to large branches and hang down to the ground, where they take root. These vertical 'cords', with a diameter of 2-10 cm, are fairly close to one another and form a kind of sleeve around the base of the tree-trunk. In a number of cases the author has observed young Allen's bushbabies playing a few metres above ground-level, leaping from one liane to another. When adults are pursued, they quite often take refuge in these characteristic aggregations of lianes. In the course of excursions through the

Figure 28. Allen's bushbaby, *Galago alleni*, in the undergrowth of primary forest. (Photo: G. Dubost)

forest, *Galago alleni* were seen relatively rarely on the ground (only 7% of observations), for the slightest alarm induces them to leap on to a narrow trunk in readiness to flee. Eight individuals were followed with radio-tracking equipment, and this confirmed that this bushbaby species seeks out forest-floor insects and fruits which have fallen to the ground. Locomotion along the ground usually involves slow progression, but sometimes leaps of 20-40 cm are made. From time to time, the bushbaby will leap on to a thin trunk at a height of less than one metre, survey its surroundings and leap back to the ground. Rapid locomotion is carried out from one vertical support to another at a few metres above ground-level, and it is usually under these conditions that Allen's bushbaby is observed. Resting periods of 15 minutes to 2 hours dispersed through the night are spent in dense foliage at a height of 10-20 metres, where observation is virtually impossible without the aid of radio-tracking to localise the resting-sites.

In Biafra, Jewell and Oates (1969) have observed this bushbaby on the ground, sometimes venturing out into small patches of grassland.

The *G. alleni* were hardly ever encountered in secondary forest, and in forest areas degraded by human activity Allen's bushbaby only occurred alongside small marshy areas (flooded patches which are not suitable for cultivation are left relatively uncleared).

(v) *The needle-clawed bushbaby*

The relatively large-bodied *Euoticus elegantulus* occupies the canopy at a height of 5-35 metres (sometimes up to 50 metres) occurring in large trees which dominate the forest. The most frequently utilised supports are large branches (49% – Fig.22), and the small claw-like extensions on the nails permit this species to move around on smooth branches of large diameter and sometimes even along large trunks. The search for gums constrains the needle-clawed bushbaby to explore particular large-calibre lianes and certain trees, particularly large trunks and branches which are inaccessible to the other prosimian species. In order to move through the canopy, *Euoticus elegantulus* may utilise lianes, but often it will run and leap along large branches in a manner reminiscent of the cercopithecine monkeys. When a needle-clawed bushbaby leaps from one tree to another, it always drops into foliage which breaks the fall (sometimes amounting to 5-6 metres).

In secondary forests, *Euoticus elegantulus* exploits the highest trees dominating the bushy vegetation in the undergrowth (*Albizia gummifera* and *Pentacletra eetveldeana*), and only on very rare occasions does it descend to the ground, where it may be surprised in the process of crossing a roadway.

The numerical data indicating the heights of supports used by the needle-clawed bushbaby (Fig.23) are a close approximation to the real situation, since this species is relatively immune to fright in the forest and often betrays its presence because of its recognisable vocalisations.

(vi) *Stratification*

An initial separation between the most closely-related prosimians can be made purely on the basis of the height above the ground at which each prosimian is typically found. Among the lorisines, the angwantibo lives in the undergrowth and the potto lives in the canopy. With the two larger-bodied galagines, Allen's bushbaby is encountered in the undergrowth and the needle-clawed bushbaby occurs in the canopy (Fig.29). Demidoff's bushbaby, which is considerably smaller than the other two bushbabies, exploits dense foliage mingled with lianes. The vertical distribution of this species thus follows the distribution of this biotope in the forest: high up in the primary forest and low down in 'tree-fall zones' or in secondary forest. With each of the five prosimian species, stratification in the forest corresponds to the search for particular biotopes where the principal dietary components can be found. The question of height relative to ground-level is not directly involved; this feature is merely a consequence of behavioural and ecological adaptations which orient these animals towards particular supports or towards specific biotopes. Thus, dietary specialisations combined with differences in spatial distribution within the forest and with differences in body-size permit these five prosimian species to coexist without engaging in interspecific competition. All of them have dietary requirements of the same general kind (insects, with a complement of fruits and/or gums for the larger forms), but each one selects from the range of suitable food-sources only certain categories for which it is specialised.

Figure 29 (right). Schematic representation of the typical pathways followed by the 5 Gabon prosimian species in primary forest. The diagram is based on a detailed vegetation plan produced by A. Hladik and C.M. Hladik (in press).

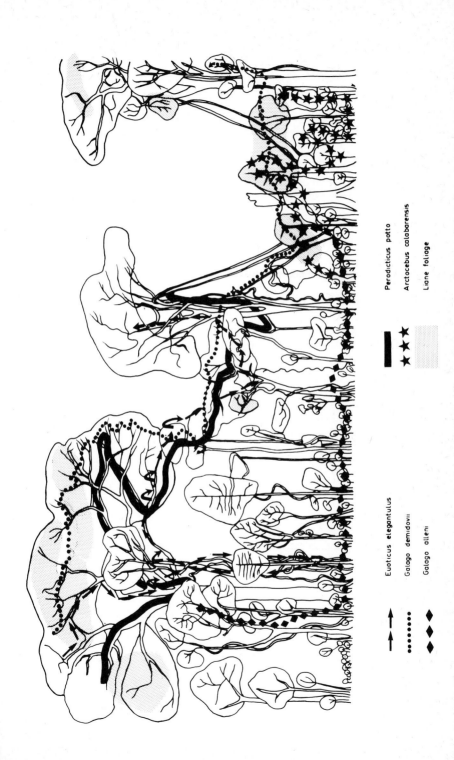

Euoticus elegantulus

Galago demidovii

Galago alleni

Perodicticus potto

Arctocebus calabarensis

Liane foliage

(vii) *Interspecific relationships*

Despite the differences in stratification and in selection of supports, there are some areas of overlap which in some cases may lead to encounters between members of the different prosimian species. Such encounters have been described in detail in a previous publication (Charles-Dominique, 1972). On no occasion, either in captivity or in the field, have direct aggressive forms of competition been observed. Under experimental conditions, if two individuals of species of different body-size are placed before a food-source at the same time, it is either the largest individual (e.g. *Euoticus elegantulus* vs. *Galago demidovii*) or the most rapid (e.g. *Galago* sp. vs. *Perodicticus*) which makes off with the food. However, such situations are highly improbable under natural conditions, given the small number of interspecific encounters and the dispersion of food-sources. Usually, two prosimians of different species which encounter one another observe each other for a few seconds and then continue on their respective ways.

In discussing the details of locomotion and defensive behaviour, one can deal with the lorisines and the galagines as two distinct groups. In fact, almost all anatomical characters which are used to justify taxonomic separation of the subfamilies Lorisinae and Galaginae are correlated with their distinctive forms of locomotion. The skull, dentition,[1] digestive tract, reproductive apparatus and other features show little major distinction; differences are only apparent in a few particular cases and the characters involved are so minor that separation at the subfamily level would not be justified on that basis alone. With the exception of the external ear of the galagines, whose role is linked to the localisation of prey, all the other major distinguishing characters of the two groups represent locomotor adaptations, diverging in two quite different directions. The Galaginae are basically adapted for leaping and rapidity of movement, whilst the Lorisinae are adapted for slow and discreet climbing locomotion.

---

[1] The extant galagines are distinguished from the lorisines primarily by the possession of a molariform last premolar in upper and lower jaws.

B. LOCOMOTION AND DEFENCE IN THE LORISINAE

The limbs of the lorisines are considerably lengthened and all the joints have great freedom of movement such that 'acrobatic' postures can be adopted. The hands and feet are broad and very muscular, whilst the well-developed thumb and big toe are markedly set off from the other digits, thus permitting a powerful pincer action. In the hand, the second digit (index finger) is greatly reduced, while in the foot a shortened second digit bears the characteristic grooming claw of the lorises and lemurs. These animals are exclusive climbers, never leaping from one support to another, and the tail is greatly reduced. In the course of locomotion any lorisine is always supported by at least two limbs on opposite sides of the body (right forelimb and left hindlimb or left forelimb and right hindlimb) and the centre of gravity is thus always located between the two points of support. This stability, combined with the power of the hands and feet, permits very slow movements during which the animal is never in a state of imbalance. The opposing limbs (e.g. right forelimb and left hindlimb) are moved forwards simultaneously, with the hand slightly in advance of the foot. At the moment where the four limbs are all in contact with the support, the hand and foot on one side of the body are in close proximity, whilst the hand and foot on the other side are widely separated. For example, the hand and foot on the right side may be in contact, whilst the hand and foot on the left are maximally separated (cf. Walker, 1969).

During their movements the lorisines do not show any pauses and the synchronisation of body actions is such that the body advances in a smooth, uniform manner. Thanks to the power of their grasp, both the potto and the angwantibo can move equally easily above or below a branch, passing smoothly from one position to the other without any break in rhythm. In this way, whenever a lorisine is moving along a branch or a liane it will pass around all obstacles (side-branch, intervening liane, chunk of dead wood, etc.), avoiding any rustling of the vegetation which might betray its presence.

Because of this slow, noiseless mode of locomotion, the lorisines avoid attracting the attention of many predators. However, since they cannot jump, they are constrained to use continuous pathways. This poses a number of problems which the potto and the angwantibo solve in different ways.

(i) *Choice of pathways*

1. *Perodicticus potto.* In the canopy zone occupied by the potto, the foliage areas of neighbouring trees are separated from one another and only overlap slightly in a few places. These scarce crossing-points represent virtually the only links between trees for the potto, and detours are obligatory. In rare situations, large lianes hang from one tree to another, providing a direct link, and the potto will use these as pathways, drawing on a thorough knowledge of the home range. In the vast majority of cases, however, the potto moves around by following branches which lead in a continuous sequence from one tree to another. In order to pass between trees along such routes, the animal descends along branches of ever-increasing diameter to the largest bifurcations and then climbs up a large branch oriented towards the junction with the neighbouring tree. To do this, the potto then has to pass along branches of gradually decreasing calibre, selecting the best direction at each bifurcation, until it reaches the area where the finest branches are mingled with those of the tree alongside (Fig.30). This behaviour explains the great variation in type of support used by *Perodicticus potto.*

When a large-calibre branch is inclined steeply upwards, the potto embraces it and advances its limbs one at a time. This form of progression permits the potto to have three points of grip at any one time. On the side where only a single grip is momentarily present, the limb involved grasps the branch at a spot intermediate between those held with the two limbs on the other side, thus ensuring greater stability (Fig.31). In addition, the potto's long, powerful digits permit grasping of branches of quite large diameter. On several occasions, the author has observed pottos attempting to reach isolated trees by descending and ascending trunks of diameters up to 60 cm. In such cases, the descent is always made with the head pointing downwards. In rare cases, the potto will venture out on to large trunks or very large-calibre branches (Fig.32), but is unable to move around with the facility exhibited by *Euoticus elegantulus.*

Figure 30. Pathway followed by a potto passing from one tree to another in secondary forest (reconstruction based on a series of photographs).

Figure 31. Young potto, *Perodicticus potto*, moving around in a bush. (Photo: A.R. Devez)

Passage from one tree to another is sometimes difficult, as the finer branches of adjacent trees are not always intermingled. Over and above this, the weight of the potto on a fine branch may cause it to bend and thus render impracticable a pathway which might seem feasible when viewed from afar. In cases of this kind, the animal will turn back and seek out a somewhat higher branch which will similarly give under its weight and thus arrive at the level of the pathway concerned. Such behaviour has frequently been observed by the author, and it would seem that the potto is aware of the flexibility or solidity of the branches along which it moves. In other cases, having sought in vain for a pathway, the potto moves out slowly to the extremity of a branch and rears up on its hind legs to reach a leaf or a small branch within range of its hands. Whatever can be grasped is then pulled in, and the foliage is progressively hauled in with the aid of the hands until a sufficiently solid branch is brought within reach (Fig.33). By subjecting the grasped branches to such tension, the potto can reach larger-calibre (and hence more resilient) branches whilst at the same time testing their

Figure 32. Potto climbing a tree-trunk by straddling it with all 4 limbs.
(Photo: M. Grange)

solidity and retaining a grasp on the first support until the
tension is sufficiently great. In the course of 100 hours of
observation of pottos in the forest, such movement from tree to
tree was seen on eight separate occasions. Each time, strong
tension was produced and the rebound of the branches when

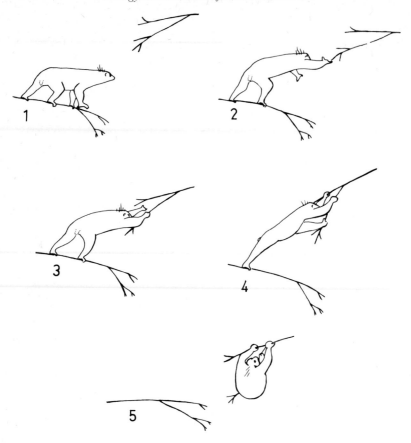

Figure 33. Potto moving between two trees separated by a small gap.

released was considerable. When abandoning the final hold
on the first branch, the potto relies on the resistance of the new
branch which has been grasped. The diameter of the second
branch is not in itself important. In fact, in captivity the
author has seen pottos grasping poorly attached branches,
pulling them in and then releasing them again despite their
large calibre. The potto is not the only animal known to pull
branches inwards by their foliage before venturing on to them.
The author has observed mandrills and chimpanzees
performing the same actions, while both the howler monkey
(Hladik, pers. comm.) and the arboreal pangolins (Pages,
1970) are reported to exhibit this behaviour.

When a potto is unable to reach a fine branch which is nevertheless relatively close by, it can perform a limited 'leap' with the hind limbs remaining firmly attached to the first branch, thus attempting to grasp the elusive adjacent branch. If the attempt fails, the animal ends up suspended upside-down by its hind limbs. This behaviour was only observed on two occasions in the field, and only in one of them did the potto succeed in reaching the neighbouring branch, which was then drawn in as with the examples cited above.[1]

Such examples of impeded passage from one tree to another are rarely observed in the wild. The potto almost always follows known itineraries. Under natural conditions, movement through the forest nevertheless requires solution of detour problems. Pottos were frequently subjected to simple detour tests, which they solved without difficulty at the first attempt. This indicates that these animals are equipped with good vision and well-developed powers of intelligence.

In the study area, where they came to visit artificial feeding sites regularly every night, pottos became habituated to the presence of human beings, but they nevertheless remained naturally cautious. They moved around fearlessly as long as the observer remained still and hidden, without any illumination, about twenty metres away. Yet as soon as they spotted the presence of the observer they moved off to hide high up in the trees. In order to observe pottos in captivity, it is equally necessary for the observer to be concealed, sometimes up to ten metres away.

Thanks to its good long-range vision, the potto selects the best itineraries in the canopy, where the arrangement of the trees sometimes necessitates long detours. In order to study a group of pottos in the wild, the author isolated six segments of the forest by cutting all of the branches permitting passage from one to another. These segments were united with movable 'bridges' composed of large-calibre lianes attached between two trees, one located in each segment. With the aid of these 'bridges' it was possible to carry out numerous observations on

---

[1] The potto never exhibits full leaps. In captivity, the author kept one for an entire year on branches attached to the ceiling in a room without a cage of any kind. Despite the fact that the distance from the branches to the ground was only 2 metres, the animal never attempted to drop to the ground to escape.

the manner in which the potto selects its itineraries through the trees.

In the first place, a potto rapidly perceives that a tree in which it finds itself is not in contact with the next tree along. If the next tree is to be visited (for example, if a male is making a visit to a female), the potto descends to the ground and climbs up the adjacent tree by scaling its trunk. If a liane has been attached between the two trees, the potto spots this at once and utilises it without hesitation, even if it is more than twenty metres in length. The position of such lianes is very important. Normally, within the forest, lianes hang beneath large branches and the potto is quite capable of selecting and finding access to one which is suitable even if, initially, it does not lead to the tree which is ultimately to be reached. Pottos were quite frequently seen making long detours, following several lianes in succession.

As an experiment, bridges connecting pairs of segments of forest in the study area were moved around. The animals were not at all disturbed by the new arrangements and immediately adopted the requisite new itineraries. Since the new bridges were constructed with fresh lianes, one can rule out the possibility that the pathways were recognised by olfactory means. At the beginning of the study, on the other hand, the bridges between the trees were constructed by throwing a cord over branches and hoisting lianes into a position above the foliage. The pottos found the lianes fairly rapidly, but they took a long time to find their ends, which were more-or-less hidden by the foliage.

In February 1967, two females were established in two segments of the study area connected by a liane which passed above the foliage of one of the trees. At the beginning, they had great difficulty in finding the end of the liane. The first evening, they made numerous fruitless attempts before finding the correct pathway. Little by little, their hesitant approaches became less and less noticeable, and after five or six days they passed directly along the liane without making any mistakes at all. Not long afterwards, the females were visited by two males. They took flight and passed across the liane to the second segment of the forest in the study area. One of the males rapidly discovered the correct pathway by following one of the females closely, but the second male hesitated a long while and did not find the pathway until the following night.

Some authors (e.g. Seitz, 1969) have put forward the hypothesis that the lorisines leave trails of urine along their pathways in order to ensure orientation within their territories. It is evident from the case cited above that the females had not marked their itinerary in this way, since one of the males hesitated for some time before finding the passage. Deposition of urine on branches under natural conditions only occurs here and there and involves quite specific spots. As with most mammals, urine-marking presumably has a territorial and social function rather than an orienting rôle. In captivity (cf. Seitz, 1969), trails of urine 1-2 metres in length are superimposed on the few branches available purely because of lack of space. It is probably this phenomenon which has led other authors to think that a potto follows itineraries marked with continuous trails of varying intensity, radiating out from a 'nucleus' where the sleeping-site is located. Pottos move around more or less everywhere within their home ranges, without necessarily returning by the same pathways each time. In general, the complexity of the canopy zone presents a broad choice of pathways, and it is only under special circumstances that the potto will use a unique itinerary. The pottos which came every evening to feed at the artificial feeding sites in the study area often hesitated before finding a crossing-point, and, apart from the bridges (with which they were well acquainted), they did not have a thorough knowledge of their pathways. The potto's memory generally permits orientation within the home range, but it is possible that scent-marks deposited at certain points may also serve as reference markers. An enormous quantity of urine would be required to lay a comparable trail along all pathways, however, since the home ranges often exceed 600 metres in diameter under climatic conditions in which rain 'washes' the branches with considerable frequency. The author has noted that under natural conditions the potto carries out urine-deposition at the same points at intervals of a few days, preferably on a large branch, where a trail of 1-2 metres is laid down.

A small group of pottos was followed over a period of three years in the study area at Makokou, and during this time the bridges between the various forest segments were frequently modified. Yet the pottos always adopted the new pathways thus imposed without hesitation, despite the fact that new lianes were used each time. The longest bridge installed had a 'Z'-shape, covering two palm-trees and crossing two roads. In order to link the two forest segments, it was necessary to attach a total length of 50 metres of lianes. Even under these conditions, and although the pottos had been crossing the ground for several months in order to move from one segment to the other, the animals immediately adopted the new pathway the first night. Although somewhat unusual in form, the new bridge was entirely appropriate to the pottos' natural habits.

2. *Arctocebus calabarensis.* The angwantibo has the same type of slow locomotion as the potto. In anatomical terms, these two prosimians differ primarily in the ratio between their sizes and body-weights. Whereas the potto is of massive, strong build, the angwantibo is slender and its weight is only 20-25% of that of the potto, despite the fact that it is only slightly smaller in actual linear dimensions (80%). The angwantibo's hands, which exhibit the same anatomy as those of the potto, are very small and only capable of grasping relatively fine supports. Living characteristically in the foliage of the forest understorey, the angwantibo frequently makes use of the fine lianes linking one bush to another, while its light build (200-300 gm) permits it to venture out on to fine twigs in search of caterpillars hidden in the foliage. Tests were carried out by placing an angwantibo on thick, vertical branches. When the diameter exceeds 4-5 cm, the angwantibo clasps its limbs around them and when the diameter exceeds 20 cm it is unable to maintain its grasp and falls off. A potto, on the other hand, can maintain a normal posture on a vertical branch 10 cm in diameter without embracing it and can climb along rough-barked trunks up to 60 cm in diameter by wrapping its limbs around the support. In fact, the two Asiatic genera in the Subfamily Lorisinae exhibit morphological distinctions of the same kind, with the slow loris (*Nycticebus*) resembling the potto and the slender loris (*Loris*) recalling the angwantibo. It is possible that the respective ecological niches occupied by these two Asiatic forms may be similar to those occupied by their African counterparts.

The undergrowth of the primary forest is relatively sparse and a climbing animal must either descend regularly to the ground or use small-calibre lianes in order to visit small trees, whose foliage often occurs as isolated patches. It is doubtless for this reason that in primary forest angwantibos are commonly restricted to tree-fall clearings where thin lianes are particularly abundant.

By keeping angwantibos under conditions of semi-freedom, the author was able to analyse their choice of pathways. A number of trees of medium height (6-12 metres in height), isolated from one another, were surrounded at their bases by cylindrical screens of smooth plastic ($1\frac{1}{2}$ metres in diameter and 70 cm in height) which prevented the animals from fleeing across the ground after descending along the trunks. In this way, three trees – each containing one angwantibo – were fully isolated.

When placed in such a tree, an angwantibo seeks to escape by gradually exploring all of the foliage. After numerous fruitless attempts, the animal will descend to the ground along the trunk, making use of a small-calibre liane wrapped around the trunk and permitting easy descent and ascent, in an attempt to find an escape route. When this proves to be unsuccessful the angwantibo climbs back up to explore the foliage and re-descends several times in an attempt to flee across the ground. Such exploration is interspersed with periods of resting and feeding and will continue for several days. Little by little, the angwantibo explores branches of increasing calibre which it had not ventured to use at the beginning. Locomotion on these branches is performed very prudently, with the animal embracing the support to obtain a better hold, and eventually the moment always comes when the animal slips and drops to the ground beyond the limit of the plastic screen. However, if thin lianes are wrapped around the large branches one can rule out such involuntary escapes. When seeking an escape route, an angwantibo explores the branches one after another to their very tips, and this is invariably followed by descent towards the ground and several minutes of movement around the plastic screen, even after several weeks under such conditions of confinement.

In order to re-establish communications between several isolated trees, bridges were set up by attaching large lianes from one tree to another. The angwantibos ventured out on to these bridges very rapidly, as soon as they passed close to their points of attachment. However, after covering 2-3 metres on these large lianes they would stop, hesitate and then turn back. Despite several attempts of this kind, they never passed from tree to tree, although it was only necessary to move a dozen metres along these lianes. This was true even when the isolated tree was linked to another which had never been occupied by an angwantibo. In fact, it emerges that the angwantibo only feels secure when sheltered amongst foliage or by some large object. This prosimian utilises small, leafy branches and lianes which pass through the foliage along tree-trunks, but it is unable to venture out on to a liane passing 'into the open'. In its natural surroundings, small-calibre lianes are common and it is rare for the foliage of bushes and small trees occupied by angwantibos to be separated by more than 2-3 metres. A liane which had previously been suspended 'in the open' was subsequently installed in the midst of cocoa trees and it was immediately utilised as a pathway by the angwantibo which had previously avoided it.

The same individual was presented with a very simple detour problem which was only solved after numerous attempts: small sections of lianes, 80 cm in length, were attached end to end to join an artificial feeding site to the branches of the tree containing the animal (Fig.34). At several points, the liane sections led to dead ends (points 3, 5 and 8), whilst at two points (A and B; indicated by dotted lines) the liane sections were replaced by fine, smooth wire on which the angwantibo would slip and could only move with difficulty.

During the first few days, the angwantibo followed the most direct path (1, A, 3, 4, B, 9 and 9, B, 4, 3, A, 1) and was thus forced to use the two lengths of wire (A, B) on which it would sometimes slither for up to 30 seconds before reaching the next section of liane. Progressively, after having explored all the liane sections, the animal adopted the pathway 1, 2,·4, 6, 7, 9 to climb up to the feeding-site and the pathway 9, 7, 6, 4, 2, 1 for the return journey. However, errors were made from time

Figure 34 (right). An angwantibo was experimentally isolated on a tree by a plastic barrier. The animal, in order to move from the feeding-site to the foliage, was obliged to use various liane segments (1-9), some of which were connected by thin, slippery wire (A,B).

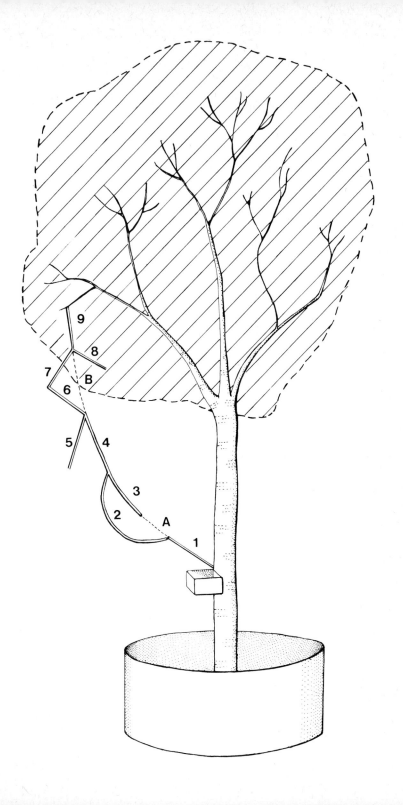

to time, particularly in the course of the descent, where the link 3 + A was often taken instead of 2. In the ascent, the detour along the curved liane (2) was learnt more rapidly than the detour involving two liane sections set at an angle (6, 7). Even after a week of such activity, when the animal was alarmed it fled upwards very rapidly and then took its former, direct pathway, which involved passage along one or two lengths of wire and a consequent loss of time. It was only when the angwantibo had calmed down that it used the detour along 2 and along 6 + 7.

It was also noted in captivity that this species had difficulty in finding pathways, even with short detours of only 50 cm to 1 metre. An angwantibo placed in an isolated tree, following a brief period of exploration of the branches, descends to the ground to reach the nearest trees. But this cannot be interpreted as a real detour, and in this particular case it probably represents a particular form of behaviour associated with a highly arboreal way of life.

It became apparent that, in contrast to the potto, the angwantibo has poor eyesight. In captivity and under conditions of semi-liberty angwantibos remain naturally cautious, but it is nevertheless possible to observe them from a distance of 1-2 metres without disturbing them, as long as the observer remains still. During observations conducted in captivity, it often happened that angwantibos would approach the immobile observer to a distance of 15 cm on a branch, abruptly stop and then turn tail. Close-range vision is also poor, and investigation of the angwantibo's hunting behaviour (see Chapter 2) showed that they are guided primarily by smell in detecting insects. On the other hand, the angwantibo is able to perceive movements easily, even from some distance away (cf. the following section on defence against predators).

The poor performance in locating detours and the weak visual powers of the angwantibo can be interpreted in terms of the animal's ecological situation. It has already been noted that this species always uses small-calibre lianes to move from bush to bush in the forest understorey. The lianes are very flexible and can only rarely have free ends like the branches and twigs of a tree. They are supported by, or attached to, rigid branches in virtually all cases and the angwantibo must thus automatically be able to reach the tree towards which it is moving at any given time, without making a detour.

Both with the potto and with the angwantibo, one can identify a close correlation between the development of intelligence, visual powers and ecological localisation.

(ii) *Defence against predators*

From the very beginning of the study, emphasis was placed on identifying the nature and importance of the predation pressures exerted upon each of the prosimian species. A large number of nocturnal predators were collected and examination of their stomach contents showed that the most common prey category consisted of rodents, whereas no remains of prosimians were ever found. The only information available for predation on the prosimian species comes from direct observation in the forest. Unfortunately, such observations are few in number, but they nevertheless provided information on the general behaviour and defence mechanisms of these animals when faced with arboreal predators under natural conditions.

In Gabon, within the study area concerned, there are nine different carnivores, six owl species and numerous snakes. Among the snakes, the python is known to prey occasionally on arboreal mammals, but no data are available on the dietary regimes of the other arboreal snakes, which are small in size. The owls have quite specialised diets. Two species feed on fish, while three others are insectivorous. (The largest insectivorous owl, *Strix woodfordi*, was frequently seen to pass close by nocturnal prosimians without attempting to capture them.) The nine carnivore species occupy well-defined ecological niches, and only a few of them may possibly prey upon the prosimians under given conditions. The leopard (*Panthera pardus*) and the golden cat (*Profelis aurata*) are almost exclusively terrestrial. They feed upon large game animals, but will also eat small mammals (rodents), and it is likely that they would attack a potto or angwantibo surprised on the ground. The same is generally true of the civet cat (*Civettictis civetta*), one mongoose (*Atilax paludinosus*) and the two genets (*Genetta servalina* and *Genetta tigrina*), which hunt on the ground and at a low-level in the bushes. Two viverrids, *Nandinia binotata* and *Poiana richardsoni*, live in the canopy, and the author has also seen (on one occasion) a black-legged mongoose (*Bdeogale nigripes*) moving around with great agility in a tree, despite the fact that this animal is normally terrestrial in habit. The diurnal raptors (eagles and hawks) hunt by sight and only discover prey animals which are on the move. The potto and angwantibo were only active during full night-time and they therefore escaped the attentions of these avian predators, which fed mainly on other birds, monkeys and squirrels.

1. *Perodicticus potto.* The predators of the undergrowth play a negligible rôle with respect to the potto, which lives almost exclusively in the canopy. Only truly arboreal predators have any real significance for this prosimian species.

The African linsang (*Poiana richardsoni*) is an arboreal carnivore of small body-size (500 gm). The stomach contents of four dissected specimens contained rodent remains and a female maintained in captivity refused to eat a dead potto provided. Only very young pottos, left by their mothers during the night, are likely to be eaten by the linsang (though this remains to be established), whilst adult pottos are sufficiently strong to resist attacks even by the African palm civet – the larger viverrid relative of the linsang.

The African palm civet (*Nandinia binotata*) is easily the most common viverrid in Gabon. The adult of this species weighs 3-5 kg and lives in the canopy, feeding on fruits, insects and (occasionally) small vertebrates. Among 30 specimens dissected, 28 had eaten fruits and some insects, one had eaten a small rat and the other had taken a squirrel. On two occasions, battles were observed between an adult potto and a young palm civet in the forest, and in both cases the carnivore eventually gave up the struggle. The adult pottos concerned remained on the defensive throughout their encounters and it is almost certain that any juvenile potto surprised under similar conditions would have been killed by the carnivore.

At 9.00 p.m. one night (15.2.67) the author observed an adult potto facing up to an immature palm civet at a height of 10 metres in secondary forest. The potto, clinging firmly to its branch, was immobile and had adopted the defence posture. Harrassed by the palm civet, which repeatedly attempted to attack from the rear, the potto switched directions time after time, maintaining a face-to-face relationship. The palm civet then attempted to attack by passing underneath the branch, but the potto thrust abruptly at the predator with its teeth and the attack from that position was abandoned. Several times, the potto projected its body forward extremely rapidly, with its mouth open, striking the branch violently with its teeth. Each 'threat' of this kind, which was accompanied by a strident vocalisation and produced a clearly audible impact between the jaws and the branch, induced the palm civet to retreat. After 15 minutes, the palm civet gave up the struggle completely and moved away. The potto remained immobile for a further 10 minutes and then continued its former passage through the tree.

At 10.00 p.m., several nights later (20.2.67) in the same study zone, an immature palm civet was seen harrassing another potto. After 5 minutes of hesitation mingled with failed attacks, the predator was knocked off balance with a violent blow from the potto's 'scapular shield' and fell to the ground. The palm civet subsequently left the scene, whilst the potto waited for 2-3 minutes before moving on.

A little while after the two observations of attempted predation, the two palm civets concerned returned to eat at the artificial feeding sites provided, yet they never tried to attack the pottos which were often present in the same trees.

The black-legged mongoose, *Bdeogale nigripes*, was only observed in the trees in one instance, where the attempt was made to capture a potto. The mongoose was stationed on the fork of a branch, 15 metres above the ground, and was attentively watching near to a young potto which the author had been following for some time. After more than an hour of such watching, in the course of which the mongoose turned its head in all directions, the potto was localised only 20 metres away, and the predator rapidly ran along the branches towards its prey. Without the author's active intervention (calling and throwing stones), the young potto would undoubtedly have been caught, but as it happened the potto moved off as the mongoose ran towards it and disappeared in a tangle of large diameter lianes. (This sequence of events was observed from a distance of 25-30 metres with the aid of a head lamp which produced reflections from the eyes of both animals concerned.)

On another occasion, a female potto which had been

released and followed close to the laboratory returned from the forest with one forelimb virtually amputated at the wrist. The hand was only attached to the fore-limb by a flap of skin and the skull was fractured in several places. These injuries were doubtless inflicted in the course of an encounter with a predatory carnivore, since fighting between pottos usually involves males rather than females and never gives rise to such serious injuries. Despite these extensive injuries, the female potto succeeded in fleeing and eventually died three days later.

These limited observations do show that the potto must defend itself against at least some nocturnal predators. Under experimental conditions, it has also been possible to study the responses of pottos faced with snakes, raptors and carnivores. However, it is mainly through observing the behaviour of pottos in the trees under natural conditions (when approached by predators or by the human observer) that analysis of their defence-systems can be achieved. These systems can only function efficiently in the natural environment, the three basic mechanisms employed being: (a) 'cryptic locomotion' and concealment, (b) active combat on branches, and (c) escape by falling under conditions of extreme danger.

(a) *'Cryptic locomotion' and concealment.* The most efficient form of protection for the potto is preventive and resides in discreet locomotion through the trees. The potto moves slowly, silently and without rustling the vegetation, maintaining a persistent, steady rhythm. It is well known that diurnal and nocturnal arboreal raptors detect their prey by their movements and by any sounds they produce. As a general rule, the mammalian eye is much more sensitive to rapid rather than slow movements, and many carnivores hunt by relying both on vision and on hearing. Olfaction certainly plays a significant rôle with some predators, but vision and hearing must be predominant in the detection of prey in the trees.

In the presence of the slightest danger or any unusual phenomenon, the potto becomes immobile and looks in the direction of the likely source of danger. Pottos were frequently surprised by the author in fairly close proximity (20 metres or so) and in all cases the animals became immobile once they had seen the approaching 'danger'. Fleeing was only evoked if the approach was continued. If, on the other hand, the author remained immobile for 1-2 minutes the potto would move away with barely perceptible movements, freezing immediately in response to any further movement by the observer.

When fleeing slowly in this way, the potto may take more than a minute to cover a distance of 50 centimetres, moving its limbs one-by-one. The further the animal moves from the source of danger, the more its speed increases, and it heads towards the densest available patch of vegetation, in which it will hide.

When there is no possibility of escape, however, the potto will turn to face the danger, adopting the defence posture. The selection of fleeing or freezing responses in fact depends to some extent on the potto's environment. If the animal is in an isolated tree at the time, it will remain completely immobile, whereas if there is a possibility for passage to another tree with denser foliage and more abundant lianes, the potto will flee there as rapidly as possible.

The potto is often obliged to use exposed pathways on which it cannot easily conceal itself. When moving under such conditions (e.g. when passing along a liane linking one tree to another, or across the ground), locomotion is rapid and the normal tempo of movement is only regained when the potto is once again under vegetational cover. As a general rule, the potto always selects those pathways where the foliage is dense, and any 'zones of insecurity' are rapidly passed by. Despite all of these complex behavioural features, however, the potto may be surprised on occasion. In such circumstances, effective defensive measures can be employed as long as the potto is on a branch or liane at the time. As with cryptic behaviour, the defensive mechanisms are only efficient in the animal's natural habitat.

(b) *Combat with predators.* A number of anatomical peculiarities of the potto are directly related to its means of defence. The apophyseal spines of the third to ninth cervical vertebrae are markedly developed; the longest protrude above the level of the skin, forming a series of 4-6 tubercles with rounded extremities (Fig.35). In this region the skin, which is

2 mm thick, is borne on the vertebral apophyses with the result that the spinal cord is located 2 cm below the surface. In the neck area, the apophyseal spines are shorter, but the skin is still 6-7 mm in thickness above the *atlas* and the *axis*. These latter two vertebrae do not bear apophyseal spines, however, since they play an essential rôle in the mobility of the head. Thus, the skin of the entire dorsal neck region is very thick, forming a collar incorporating an epidermis of special structure. The surface is covered with alveoli (pits) containing bunches of bristles, some of which protrude beyond the fur (which is just as dense in this zone as on the rest of the body surface). Histological studies conducted by Montagna and Ellis (1959, 1960) have shown that these long bristles, which are clearly distinguished from vibrissae (whiskers), represent a particular type of sensory hair. The same studies also demonstrated that the tubercles are rich in tactile corpuscles, which are just as abundant as on the lips or on the penis.

The tubercles, in particular, have attracted the attention of numerous authors and these intriguing structures have been interpreted in a variety of ways. They have been identified as components of a defensive system (Sanderson, 1937, 1940; Cansdale, 1946; Rahm, 1960), as protecting devices for the vertebral column (Eimerl and de Vore, 1965), as an abrasive organ (Cowgill, 1969), as adaptations for interspecific combat (Blackwell and Menzies, 1968), as sites of muscular attachment (Jewell and Oates, 1969) and as sensitive zones involved in social contact behaviour (Walker, 1968, 1970).

Following the author's field observations of combat between pottos and carnivores, it is evident that the neck region is quite definitely involved in the defensive behaviour of the potto. In fact, it is not sufficient to consider the tubercles alone; one must take into account the entire set of anatomical

Figure 35 (right). The scapular shield of the potto.
A. View from above, indicating the thick neck region covered with alveoli containing long tactile hairs (exposed by shaving the hair).
B. Lateral view, showing the tubercules. Note the position of the shoulder-blades (scapulae), which partially cover the vertebral column.
C. Sagittal section of the same region of the scapular shield. The apophyseal spines of the vertebrae, together with the thickening of the skin and the position of the scapulae, combine to protect the vertebral column situated 2 cm below the surface. (Diagram prepared from a photograph by Catherine Muñoz-Cuevas)

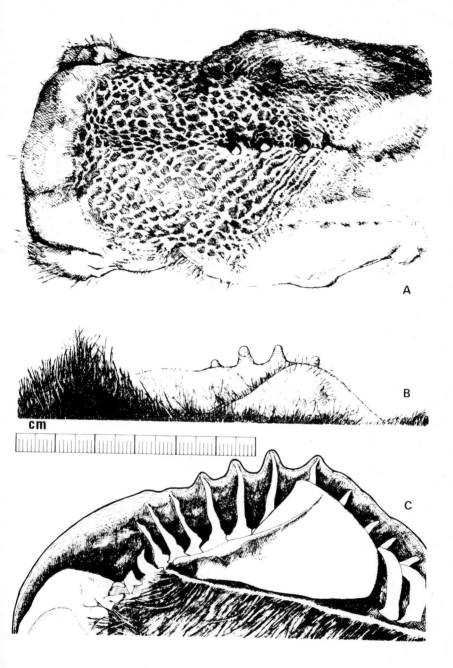

cm

A

B

C

peculiarities involved in the overall structure referred to here as the 'scapular shield': the dermal collar of the dorsal neck region; the long apophyseal spines of the vertebrae (which form a veritable brace in the dermis); and the peculiar dorsal position of the scapulae of the potto, ensuring that even part of the vertebral column is covered.

This composite, but nonetheless effective, shield in no way hinders the potto and the neck remains relatively supple despite its protective covering. With one potto, which survived for three days following an encounter with a carnivore in the forest (see above), transverse sections of the neck region (Fig.36) showed that bites had been inflicted near the 4th and 6th vertebrae, but they had not penetrated through to the spinal cord. This protection is further enhanced by the great tactile sensitivity of the scapular shield which plays a significant sensory rôle at the moment when the potto is hiding its head and is hence prevented from seeing its adversary (Fig.37A).

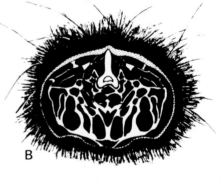

cm |ᵖᵖᵖᵖᵖᵖᵖᵖᵖᵖᵖᵖᵖᵖᵖᵖᵖᵖᵖᵖᵖᵖᵖᵖᵖᵖᵖᵖᵖᵖᵖᵖᵖᵖᵖᵖ|

Figure 36. Transverse sections through the neck of a potto.
A. Section at the level of the 4th vertebra.
B. Section at the level of the 6th vertebra.
C. For comparison, a transverse section of the neck of an adult angwantibo.
(Drawings prepared from photographs by Catherine Muñoz-Cuevas)

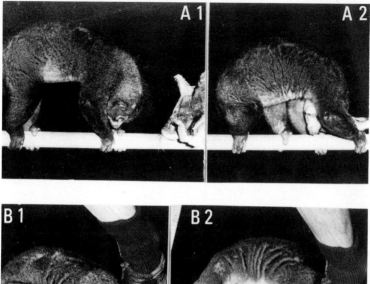

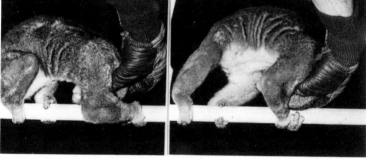

Figure 37. Defensive posture of the potto.

A. When threatened by some kind of danger (the skin of a carnivore in this case), the potto protects its head between its forelimbs, presenting its scapular shield to the adversary. At this point, the potto is no longer able to see its adversary, and sensory information is provided by the tactile hairs.

B. When seized by the neck, the potto thrusts vigorously such that, under normal conditions, the adversary may be forced to release its grip and may even be toppled to the ground.

In the defence posture, the forelimbs are tensed whilst the hind-limbs are semi-flexed and the hands and feet are brought together. The head is lowered such that the adversary is presented only with the region of the head and shoulders protected by the scapular shield, but the potto is still able to observe its opponent. When attacked by a carnivore, the potto protects its head even more by lowering it still further between its arms. Without releasing the grip of its hands, it can then swing its body backwards or sideways to avoid the adversary. If grasped by the neck, the potto repeatedly jerks sharply forwards and this movement can force the aggressor to lose its hold. Sometimes, immediately after avoiding an attack by swinging its body away, a potto may make a violent thrust with the scapular shield[1] (see p.85), which can be sufficient to knock a carnivore off balance and even to dislodge it from the branch (Fig.37B). Carnivores cannot cling so firmly to the branches as can the potto, thanks to its very muscular hands. As a test of this tenacity of the potto, the author once attached a weight of 15 kg to a potto without breaking its grip on the branch. (It should be noted that Suckling et al., 1969, have demonstrated that the peculiar circulatory system of the potto's limbs – involving an arterial *rete mirabile* surrounding the veins – permits continuation of the venous circulation even when prolonged muscular contraction occurs.)

In the fully retracted posture the potto is no longer able to see what is happening directly ahead, and it is here that the long bristles and the tactile corpuscles of the tubercles take over the predominant sensory rôle. Under experimental conditions, this behaviour can be evoked with a dummy[2] (see p.94). When the potto has accordingly adopted the retracted posture, the thrusting movements can be elicited by brushing the tactile hairs or by grasping the neck (Fig.37B). Thus, the scapular shield has a double function: it is both sensory and protective. As a general rule, adaptation of an organ for a protective function need not necessarily be accompanied by a reduction in sensory function. For example, in otters the extremely thick lips – which serve as a protection against fish-spines – maintain great tactile sensitivity.

---

[1] These thrusts of the scapular shield have been described by Sanderson (1937, 1946), by Rahm (1960) and by Blackwell and Menzies (1968), who have all interpreted this movement as an offensive gesture associated with the presence of the tubercles. The first author had not adequately examined the tubercles, since he described them as consisting of bony points projecting directly through the skin. Rahm, on the other hand, remained sceptical with respect to this suggested function, having observed that the ends of the tubercles are rounded and (of course) covered by a layer of skin.

[2] The dummies used in these experiments were skins of three carnivores: *Genetta servalina*, *Nandinia binotata* and *Profelis aurata*. The potto's responses are usually most characteristic during the first few minutes following presentation of the dummy. After several presentations of the dummy, the potto's attention becomes directed mainly towards the observer rather than to the carnivore skin.

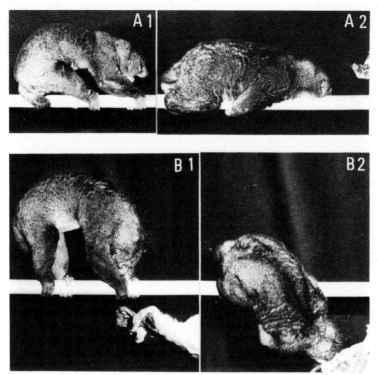

Figure 38. A potto attacking a dummy (carnivore skin).

A. Without releasing its grasp on the support, the potto thrusts its body forwards and directs a violent bite at the support whilst uttering a hoarse vocalisation (see sonogram in Fig.69).

B. When attacked from beneath, the potto aims a bite downwards.

Walker (1968, 1970) has expressed the view that the sensory zone of the neck region is involved in social contacts as a 'tactile focus for peaceable interactions in the species'. *A priori*, there is no reason why such a function should be incompatible with the protective rôle described above. However, Walker's hypothesis can only be confirmed or invalidated by a detailed investigation of the social behaviour of the potto.

The scapular shield is not the only defensive means available to the potto when faced with a carnivore. Rahm (1960) has described the manner in which a potto may bite its aggressor, and the author has frequently observed such offensive behaviour (in response to African palm civets or to a dummy presented under experimental conditions). Without releasing its grip on the branch, the potto thrusts its body forwards with its mouth open and utters a hoarse vocalisation (Fig.38A). The same pattern may be exhibited when biting an aggressor located beneath the branch (Fig.38B), and in either case the potto immediately recoils to its initial posture. The potto possesses stout canines and premolars, whilst its masseter muscles are very well developed, and any bite on an adversary is usually maintained for a certain time before the grip is released. This pattern of attack is only utilised in response to carnivores; in intraspecific fights, the hands are used as well.

As stated above, the potto is only able to defend itself effectively in this way when on a branch, where the adversary must approach from one direction and cannot move around to vary its attack. On the ground, this no longer applies, and the author has seen a potto killed by a dog when exposed in this way. In the trees, however, whenever the predator moves through the branches to attack from another direction, the potto switches its stance and maintains its defensive posture.

The potto doubtless falls victim to arboreal carnivores only on rare occasions. As has already been seen, even a female which had been fatally injured was able to escape from her adversary, thus rendering the combat fruitless. It seems likely that the characteristic strong, curry-like odour of the potto's fur may serve as a warning signal to carnivores which have already been repulsed in a combat. (As has been seen, young African palm civets, after an unsuccessful struggle with a potto, will not try to attack one in any subsequent encounter.)

This complex of defensive behaviour first appears at about the age of 2 months in young pottos engaged in play behaviour, usually involving the mother. It is common to see a young potto in the defence posture riding on the back of its mother whilst she is moving through the trees. Whilst on the move in this way, the juvenile may thrust its scapular shield or strike with its teeth at a passing leaf. Sometimes, the young potto bites its mother's ears, and she turns round to utter an aggressive vocalisation. On two occasions, a young potto 'parked' in a tree by its mother was seen to simulate combat with a small branch, which was shaken violently by the thrusts.

(c) *Falling to escape from a predator.* When extremely frightened, a potto may simply fall to the ground. Most individuals (7 out of 8) seen to behave in this way were responding to a large snake (a dead cobra) which was presented experimentally. The eighth potto retained its grasp on the branch, but did not attempt to bite the snake. Some of the animals tested were so frightened by the experience that they let go once again when replaced on the branch. It should be noted that this same fright response was exhibited by one potto when presented with a genet skin. In all other cases, this dummy provoked combat behaviour, as described above.

On one occasion, when the author was about to capture a female potto in a tree, the juvenile she was carrying on her belly simply dropped to the ground. The young potto remained immobile a few metres away from the point of impact, yet it took half-an-hour for the author to find it. It was also observed in one instance that a young potto dropped to the ground when a rifle shot was fired nearby.

Falling from the trees in this way reflects the highest degree of fright in response to an adversary. Film analysis (Fig.39) shows that the animal abruptly recoils and then adopts the defensive posture, whilst at the same time releasing its hold on the branch. After striking the ground, the potto rapidly moves a distance of 2-3 metres and then remains immobile.

The overall predation pressure exerted on populations of the potto must be very limited. The females have only one infant a year and sexual maturity is only reached at about the age of two years. Predation is certainly not the only cause of mortality under natural conditions. Pottos in the wild may suffer from various diseases.[1] Two pottos (one juvenile and one adult) were captured in a moribund condition, and a third was found dead in the forest without any trace of injury. All of the pottos captured in the forest were parasitised by a large number of nematode worms (*Filaria; Ascaris*), by ankylostomes and by *Hymenolepis diminuta*. The *Ascaris* can be so numerous as to form a bulky package occupying most of the volume of the stomach cavity.

---

[1] One of the diseases observed was expressed by loss of weight accompanied by nervous disorders (trembling and paralysis beginning at the hind end of the body).

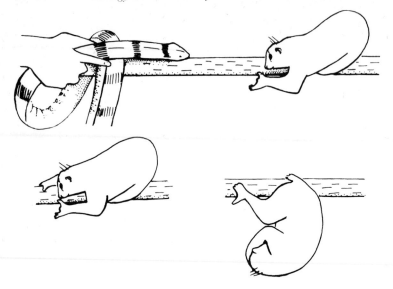

Figure 39. A potto falling after presentation of a snake. (Illustration prepared from a film strip)

2. *Arctocebus calabarensis.* Since this species lives in the undergrowth and frequently descends to the ground, it is doubtless more exposed than the potto to attacks from terrestrial and semi-arboreal predators, which are quite common (viverrids, felids, snakes, etc.). However, no data are available concerning the natural predators of the angwantibo, though some observations of defensive behaviour were conducted in the forest in cases where the author's presence represented the 'danger' evoking the response.

(a) *Slow, noiseless locomotion and concealment.* The angwantibo has the same mode of locomotion as the potto, and this doubtless provides the principal form of protection against predators. In the presence of danger or any unusual phenomenon, the angwantibo freezes and remains immobile. Most of the individuals encountered in the forest were already immobile. When in this state, they can be approached closely as long as the observer moves slowly, silently and (above all) without touching any lianes or small trees.

At the slightest rustling of the vegetation, the angwantibo flees upwards and freezes again at a height of 5-10 metres. The animal never attempts to face the source of danger and seeks continually to flee upwards, even if the pursuer is only a few centimetres away. When fleeing, the angwantibo climbs rapidly up small-diameter vertical lianes, along which viverrids have more difficulty in moving. Despite its poor sight, *Arctocebus* exhibits fine perception of movement even over great distances. One individual was seen to flee rapidly into the foliage of a tree when an insectivorous owl (*Strix woodfordi*) glided past more than 10 metres away.

(b) *Defence posture.* If an angwantibo is grasped violently, it remains attached to the branch with its body rolled into a ball. On the other hand, if it is not seized at once it may fall to the ground before capture can be effected. As with the potto, there are quite marked individual variations in this response.

The defence posture is somewhat unusual. The head is completely retracted between the arms, but the mouth is held open and directed beneath one of the arm-pits (cf. Jewell and Oates, 1969). At the other end of the body, the hairs on the tail have black tips. When the animal is undisturbed, all that one can see at the end of the tail is a small black patch. However, as an effect of fear the hairs are raised to form a dark circle surrounding a pale zone formed by the tail tuft. In the defence posture, only the tail is conspicuous since the head is concealed beneath the body (Figs. 40, 41). As soon as an angwantibo is touched anywhere at the rear of the body, one arm is lifted and the adversary is bitten as the head passes beneath the arm-pit. Every untrained observer who has attempted to seize an angwantibo has been bitten in this way. In addition, since the animal does not immediately release the clamp of its jaws, it is propelled a distance of several metres by the abrupt recoil movement of the hand. When it is appreciated that the angwantibo descends fairly frequently to the ground and that it is there, above all, that it is in danger of being killed, the effectiveness of such a defence mechanism can be readily understood. Since the head is kept out of sight, the dark circle formed by the raised tail hairs may attract the attention of a carnivore,[1] and it is highly likely that under

[1] It is possible that other factors (olfactory in nature) may orient the attention of predators towards the tail. When an angwantibo corpse was presented to a genet, it was noted that the carnivore first attacked its 'quarry' from the rear.

these conditions the predator would be bitten on the snout, provoking an abrupt recoil movement. As a result, just as in the case where the human hand is bitten, the angwantibo could well be thrown some distance and might subsequently be able to make its escape. However, this hypothesis remains to be substantiated by further field observations.

Figure 40. Defence posture of the angwantibo, seen from one side (*A*). The head is concealed beneath one armpit, whilst erection of the tail-hairs brings out the pale-coloured tuft of hairs surrounded by a black ring (*B*). In this position, the angwantibo can bite by raising its arm and passing its head beneath the armpit, if it is grasped from above or from behind.

A

B

The defence postures and mechanisms of the potto and the angwantibo do differ significantly. With the potto, the scapular shield and the biting attack are only effective when directed against a predator located directly ahead, on the same branch. With the angwantibo, the protective device is only effective when the animal is on the ground, with the effect directed upwards towards a predator attacking from above. The defence mechanisms of the two species have no doubt evolved as a function of the most frequent predator-exposure situations encountered.

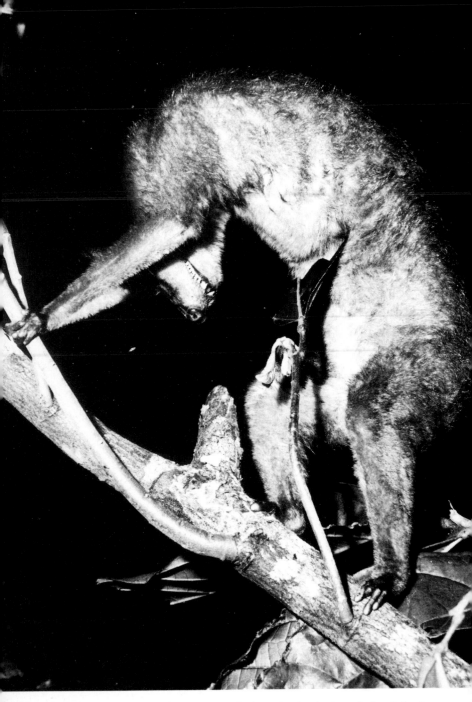

Figure 41. Angwantibo photographed at the beginning of the defensive posture. (Photo: A.R. Devez)

C. LOCOMOTION AND DEFENCE IN THE GALAGINAE

(i) *Prehension and branch-running locomotion*

Adaptation for leaping no doubt constitutes the most striking feature of the galagines, but running and the mode of prehension nevertheless play an important part in the biology of this prosimian group.

Several authors (cf. Bishop, 1964) have demonstrated that the prosimians have only poor neuro-muscular control over their hands. The fingers are long and it is the terminal phalanges, each of which has a very well developed tactile pad, which enter into contact with any objects which are grasped. There is characteristic flexion of the fingers at the articulations between the 2nd and 3rd phalanges (Fig.42). When running along a branch, any bushbaby maintains a grasp on the support with at least one extremity at any given time. With the foot, the grasp is always maintained by a pincer-action between the hallux (big toe) and the other toes. By contrast, the pollex (thumb) of the hand is not clearly opposed to the other fingers, which are held in a fanned-out arrangement with the terminal phalange contacting the support as a result of the characteristic pattern of flexion of the fingers. In the resting position, or when the hand is placed on a plane horizontal surface, the hand phalanges are moderately flexed. It is only on large branches that the fingers are fully extended so that the hand can cover a maximal surface area.

With the needle-clawed bushbaby, as was seen in Chapter 2, there is a special adaptation to permit climbing on smooth trunks and large branches in the course of searching for gums. The nails on digits 2, 3, 4 and 5 of the hand and on digits 3, 4 and 5 of the foot are keeled, and the keels are all prolonged to form a small 'claw'. The tactile pads of the terminal phalanges are markedly expanded so that in the resting position the 'claws' do not touch the support. It is only when the terminal phalanges are mildly flexed downwards that the tips of the nails contact the support, with the result that the supple tactile pads are pushed backwards by the pressure exerted on the fingers (Fig.43). When moving over smooth supports of large diameter, *Euoticus elegantulus* spreads its limbs widely and

Figure 42. Demidoff's bushbaby photographed whilst running. Note the position of the digits of the left hand, which are flexed and only contact the support through the terminal phalanges. (Photo taken at 1/50,000th of a second by A.R. Devez)

moves them in the following order: right fore-limb, left hind-limb, left fore-limb, right hind-limb. This sequence is the same as that exhibited by the lorisines when they are embracing large branches, and the needle-clawed bushbaby is able to move around with facility in this way, sometimes with the head directed downwards (Fig.44). Apart from this highly specialised form of locomotion, *Euoticus elegantulus* leaps and runs on large branches just like the other bushbabies. However, special adaptation for movement on supports of large diameter is reflected by the fact that this species has, in proportion to its body-size, the largest hands and feet and the most highly developed tactile pads on the terminal phalanges.

The branch-running locomotion of *Galago demidovii* has been studied with the aid of analysis of film and single photographs (Fig.45). The hind-limbs exert the greatest effort in propelling the body forwards, whilst the fore-limbs are primarily concerned with support and stabilisation since the body is continuously disequilibriated during branch-running, which is often interspersed with short hops. The tail is also involved in the maintenance of balance.

It is noteworthy (Fig.45) that in *Galago demidovii* the elongation of the tarsal region of the foot considerably increases the hind-limb stride length without hindering retraction of the limb, which folds into three segments of equivalent length. In *Euoticus elegantulus*, which is an excellent leaper, the tarsus is relatively smaller than that of the species of the genus *Galago* (Martin, 1972b) and it is the first two segments of the hind-limb (femoral and tibial segments) which are considerably elongated. 'True running' of the

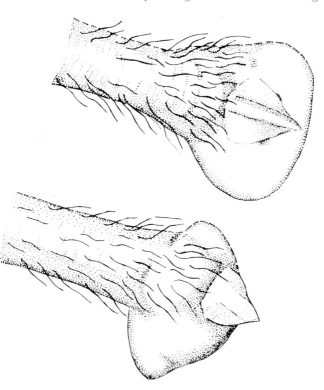

A

B

Figure 43. Digit of the hand of a needle-clawed bushbaby.
A. Usual posture – the tactile pad adheres to the support, and the 'claw' is not in contact.
B. When the terminal phalange is retracted, the tactile pad of the digit is deformed and the 'claw' digs into the support.
(Drawings prepared from original photographs)

Figure 44. Movement of a needle-clawed bushbaby on a tree-trunk. The limbs are spread and moved one-by-one, whilst a firm hold is maintained on the support with the 'claws' at the tips of the nails. In this posture, the bushbaby can easily descend along the trunk, with its head pointing downwards. (Drawing prepared from a film-strip)

needle-clawed bushbaby is relatively slow. When the hind-limb is brought forward, the animal is obliged to swing the foot outwards somewhat whilst raising the rear end of its body. Rapid locomotion is essentially achieved by a sequence of short leaps, a pattern which is entirely feasible on the relatively unencumbered large branches usually used by this bushbaby species.

Adaptation for leaping requires elongation of the leg bones and this presents difficulties for rapid limb retraction in running. Thus, leaping and running are to a certain extent incompatible. Within the genus *Galago*, this elongation of the hind-limb is primarily expressed in the tarsus and this permits rapid forward movement of the leg during running. Thus it would seem that, at least within the subfamily Galaginae, development of the tarsus constitutes an adaptation permitting a combination of leaping and running. Indeed, *Euoticus elegantulus* is the most effective leaping form, whilst *Galago demidovii* – which has the best developed tarsus relative to its size – is the best adapted for running. *Galago alleni*, whose

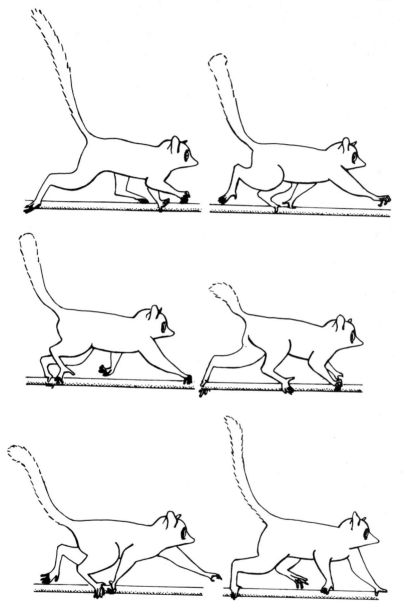

Figure 45. Schematic sequence of running locomotion of Demidoff's bushbaby. (Reconstructed from individual photographs)

tarsus is also well developed, exhibits great prowess in both running and leaping. This interpretation would at least appear to hold good within the subfamily Galaginae. However, there are a number of different 'solutions' to the problem of adaptation for leaping, according to the ecological pressures experienced, and any broad generalisation on this subject is hence somewhat suspect.

(ii) *Leaping*

Bushbabies have extremely well-developed vision, which is an indispensable requirement for an arboreal leaping form. Studies conducted on *Galago senegalensis* have shown that this species possesses great visual acuity and well-established depth perception (distinction of 7.7 lines per cm at 1 metre and perception of 1 cm depth of field at 1 metre; cf. Treff, 1967). The three bushbaby species studied in Gabon also seem to possess finely developed visual powers. They have been observed leaping with great precision in the trees even on quite dark nights (though under conditions of extreme darkness their activity is notably reduced). In the bushbabies generally, the great development of the eyes would itself seem to be correlated with specialisation for leaping locomotion. It is also important to note that the tail, which is well-developed in the Galaginae, performs a significant balancing function in the course of the leap.

Bushbabies perform leaps over distances which are considerable in relation to the animals' body-size, but certain authors have nevertheless exaggerated their actual achievements. In the course of observations in the field, the distances covered by leaping (without loss in height) were measured on several occasions for each species. (N.B. In captivity, bushbabies are often in a weakened condition, and their leaps are accordingly less extensive.) *Galago demidovii* can leap to a branch 1.5-2 metres away without losing height, whilst *Galago alleni* and *Euoticus elegantulus* can both cover 2.5 metres under the same conditions.

In the course of general locomotion, the bushbaby species exhibit running and leaping in roughly equal proportions (with the exception of Allen's bushbaby, which constitutes a special case). This mode of locomotion is quite different from that of the squirrels (for example), which typically run along branches and exhibit leaping largely in order to attain the foliage of a neighbouring tree, which acts to soften the impact of landing. The bushbabies, by contrast, often leap from branch to branch within one and the same tree. This permits very rapid movement, but it necessitates a mechanism to absorb the shock of landing on rigid supports. The supports utilised are often vertically oriented, and it is interesting to note that Napier and Walker (1967) consider 'vertical-clinging-and-leaping' to have been a common adaptation among early Tertiary primates.

Each bushbaby species studied in Gabon exhibited a specific pattern of locomotion. The differences between the species are doubtless dependent on differences in adult body-weight and in the ecological localisations characterising each species.

1. *Galago demidovii*. This bushbaby species is characterised by its small body-size (head + body length = 12 cm) and by its low body weight (average 60 gm). It typically occupies dense vegetation, where the supports utilised are fine and oriented in diverse directions (Figs. 24 and 25). The trajectories followed in leaping are virtually straight so that both in leaping to vertical branches and in landing on horizontal supports the shock of impact is taken up by the fore-limbs (Figs. 46 and 47). When leaping to a vertical branch, the animal lifts its head backwards just before contacting the support. The fore-limbs suffice to arrest the movement of the body and the hind-limbs only touch the support at the last moment, when the body resumes its vertical posture. *Galago demidovii* may complete its leaping movement in this latter position, subsequently switching to running along the support, or it may swing around the branch (if it is of sufficiently small

Figure 46. Analysis of leaping between horizontal supports in Demidoff's bushbaby. The forelimbs absorb the shock of landing. (Illustration based on a film-strip)

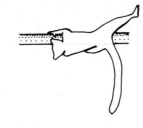

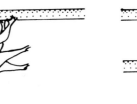

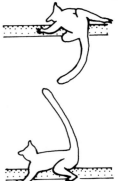

Figure 47. Analysis of leaping between vertical supports in Demidoff's bushbaby involving rotation around each support. The forelimbs absorb the shock of landing and the tail is primarily involved in re-establishing balance after take-off in the leap. (Reconstruction from a film-strip)

diameter) to leap off again. Before taking off, however, the animal retracts its body and hesitates for a short period of time.

At the beginning of the leap, the body is twisted around its long axis (Fig. 47), and balance is regained in mid-leap by a whipping movement of the tail, whilst the body becomes horizontal. On landing followed by pivoting around the support, it is the tail which swings the body laterally.

2. *Galago alleni.* This bushbaby weighs four times as much as *Galago demidovii* (viz. 250 gm on average), and its locomotion is the most specialised. The species moves almost exclusively by leaping between thin trunks and the bases of lianes, passing at great speed from one vertical support to another and disappearing from view in the space of a few seconds. In order to achieve this, Allen's bushbaby pivots rapidly around each support, taking off almost immediately to leap on to the next. Analysis of film and single photographs (Figs. 48 and 49) has shown that only half a second elapses between the movement of impact on a thin trunk and the instant of take-off. In five seconds, *Galago alleni* can easily cover a dozen metres in five or six leaps of this kind.

At night it is easy to identify *Galago alleni*, even when it is only possible to see the eyes, from this rapid pattern of leaping and the very brief intervals between successive leaps. The trajectories of the leaps performed by Allen's bushbaby are very flat (though not as straight as in *Galago demidovii*). At the beginning of the leap, the body is fully extended and almost

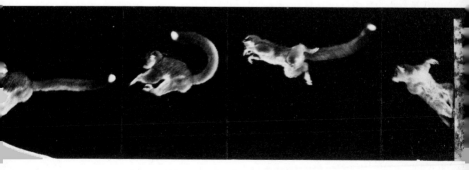

Figure 48. Photographical analysis of leaping between vertical supports in Allen's bushbaby. (Photos: A.R. Devez)

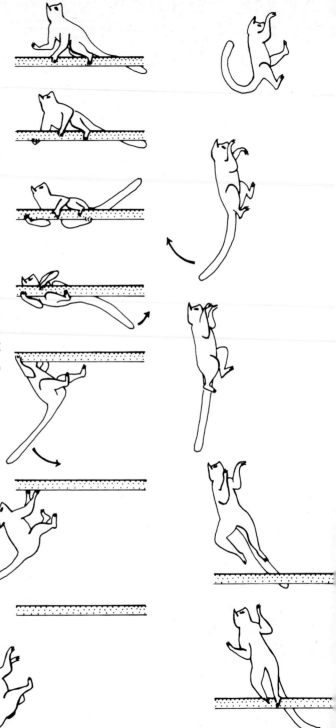

Figure 49. Analysis of leaping between vertical supports in Allen's bushbaby, with rotation around the support. Before landing, the tail is raised onto the back, and it aids in returning the animal's body to a vertical alignment by means of a whip like thrust downwards. Even before the support is reached, the head is already turned towards the subsequent support. This prior orientation of the head is maintained throughout the entire phase of rotation around the intervening support. (Reconstruction from a film-strip)

horizontal, which permits rapid movement through the undergrowth despite the presence of lianes and bushes. The fore-limbs play only a secondary rôle in the absorption of the landing impact and most of the kinetic energy is taken up by the hind-limbs and by rotation around the support. The hind-limbs are brought forward in the course of the leap, but not as far as with the tarsier, the sportive lemur, the indriids and *Galago senegalensis* (Petter, 1962; Walker, 1967; Bearder and Doyle, 1974).

At the moment when the hind-limbs are swung forwards, the tail is brought up to arch over the back and is subsequently whipped rapidly downwards when the hands contact the support. This 'whip-lash' movement contributes to the vertical alignment of the body, whilst the backthrust exerted by the fore-limbs on the support operates in the same sense.

Carried on by its momentum, the bushbaby pivots around the vertical support and – following very brief retraction on to its hind-limbs – takes off in another leap. The thin trunks or liane bases used in such locomotion are not necessarily aligned, but following each leap the animal arrives at the optimal position (i.e. with its back turned directly towards the next support) before taking off again. In other words, Allen's bushbaby controls its angle of rotation, following each landing, in accordance with the pathway followed through the undergrowth.

When the pivoting movement commences, the head is already turned towards the next vertical support. This orientation of the head is consistently maintained throughout the rotation movement around the intervening support, with the result that – after pivoting to the required angle – the head is rotated at 180° relative to the body just prior to renewed take-off. The angle of rotation of the head relative to the body undoubtedly plays a part in control of the pivoting movement, thus enabling *Galago alleni* to adopt the optimal positions prior to each successive leap. The two other *Galago* species in Gabon do not exhibit such an adaptation for leaping on vertical supports.

The form of play behaviour preferred by young *Galago alleni* is that of leaping among vertical supports and pivoting around them, sometimes through an angle of 360°. This 'conditioning', in addition to the specific preference that the young animals exhibit for vertical supports from the age of one month onwards, provides a good demonstration of the degree to which this species is adapted for leaping between vertical supports. This would seem to have a definite genetic basis, as is indicated by observations conducted with two young *Galago alleni* captured when they were still incapable of independent locomotion (ages: 3 weeks and 4 days respectively) and subsequently reared in captivity. As soon as they began to leap, despite their freedom to move anywhere in the house, these two animals spent their time moving among the vertical legs of the chairs and tables. They leapt from one to the other with great agility, playing in this fashion for long periods of time. Young *Galago demidovii* and *Euoticus elegantulus* reared under the same conditions never exhibited such attraction to vertical supports.

3. *Euoticus elegantulus*. The needle-clawed bushbaby is of the same general size as Allen's bushbaby (head + body length = 20 cm), but it is somewhat heavier (average: 300 gm). As has been seen, it is exclusively active in the forest canopy, moving over large trees whose broad branches are usually oblique or horizontal in orientation (Fig.25).

Whatever the orientation of the support terminating the leap (vertical, oblique or horizontal), *Euoticus elegantulus* always follows a trajectory with a sharp upward curve and with the body held erect throughout. When the point of arrival is an oblique or vertical support, the four limbs take up the landing impact simultaneously (Figs. 50 and 51), but the animal never pivots around the support in the manner exhibited by *G. demidovii* and *G. alleni*. The supports used are generally of large diameter and it is therefore impossible for the needle-clawed bushbaby to establish a complete grasp with the hands and feet. However, as mentioned above, the hands are very large and the tactile pads of the terminal phalanges are particularly well-developed, and this permits effective adhesion to large-calibre branches. (N.B. The

Figure 50. Needle-clawed bushbaby leaping on to a vertical support. The animal's body remained vertical throughout the leap. (Photo: A.R. Devez)

pointed tips of the nails are only used in climbing, not in leaping.) Leaping along a sweeping trajectory requires less energy than that conducted with a flat trajectory (see Charles-Dominique and Hladik, 1971). The shock of landing is accordingly less violent, though such leaping behaviour is only possible in surroundings where there is little obstruction between supports.

Euoticus elegantulus frequently moves from one tree to another by executing leaps over considerable distances, always in a downward direction and involving a loss in height which may be as great as 8 metres. In this way, this bushbaby species can move between widely-spaced supports, and one leap measured in the field was found to involve a horizontal displacement of 5.5 metres and a loss in height of 3.1 metres. On landing, a grasp is made on available foliage, which softens the impact of arrival. In the course of free fall (Fig.52),

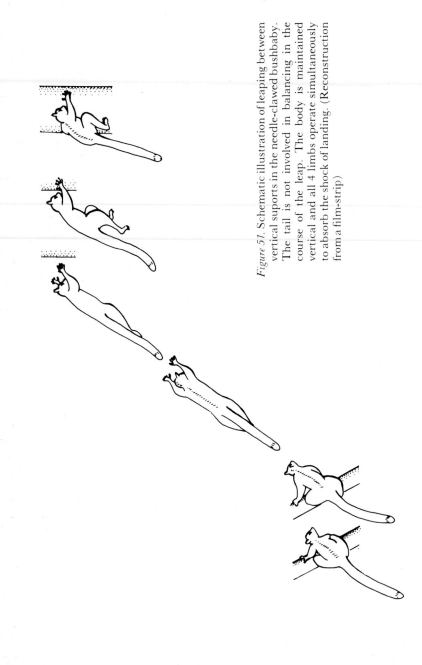

Figure 51. Schematic illustration of leaping between vertical suports in the needle-clawed bushbaby. The tail is not involved in balancing in the course of the leap. The body is maintained vertical and all 4 limbs operate simultaneously to absorb the shock of landing. (Reconstruction from a film-strip)

the widely spread limbs – which present a maximal surface area[1] – and the bushy tail act to decelerate the movement. *Galago demidovii* sometimes leaps in this fashion in order to reach the foliage of a neighbouring tree, but this species has never been observed to fall more than 1.5 metres in the course of a leap. It is interesting to compare this behaviour with the initial phases of leaping in *Galago senegalensis* (Hall-Craggs, 1965, 1974) and in *Galago alleni* (Jouffroy, 1975; Jouffroy et al., 1974).

(iii) *Defence against predators*

No data are available with respect to predation upon the three bushbaby species in Gabon. Encounters between Demidoff's bushbabies and *Nandinia* were observed on several occasions, but the carnivores seemed to be indifferent to the bushbabies, which harrassed them and followed them at a distance of several metres whilst uttering intense alarm calls. Struhsaker (1970) also observed *Galago demidovii* producing alarm calls in the proximity of an arboreal snake. This harrassing response ('mobbing') is quite common in the forest, and it has also been observed with the other two bushbaby species. It seems to be evoked by any dangerous or 'strange' animals.[2] *Euoticus elegantulus* was observed harrassing an arboreal pangolin (*Manis tricuspis*) in this way, and all three bushbaby species may exhibit this response in the presence of human beings. The local Gabonese villagers also maintain that *Galago alleni* responds in this way to the presence of a leopard, and this is hardly surprising in view of the fact that Allen's bushbaby inhabits the undergrowth zone. This harrassing response doubtless has a value as a social warning system, for when such calls are heard conspecifics become more cautious and they may sometimes even discreetly approach the individual uttering the alarm calls. In the latter case, it is very rare for the

[1] In the needle-clawed bushbaby there is a skinfold at the armpit and the junction between the thigh and the trunk, giving the appearance of an incipient *patagium* (gliding membrane) when the limbs are spread during the fall.

[2] However, it should be noted that the owl *Strix woodfordi* was observed on numerous occasions hunting insects a few metres away from *Euoticus elegantulus*. Unlike the angwantibo, the needle-clawed bushbaby did not respond to the owl's presence.

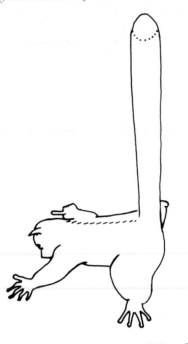

Figure 52. Illustration of a needle-clawed bushbaby falling. In order to move from one tree to another, this bushbaby can leap downwards to grasp foliage located up to 8 metres below the point of departure.

newcomer to participate in the harrassment behaviour. The calls may also exert an effect between species; individual *Galago demidovii* and *Euoticus elegantulus* have been observed to approach an Allen's bushbaby engaged in uttering alarm calls in the wild.

The bushbabies can escape from any predator they notice by leaping away. Only individuals which are approached by surprise are likely to be killed, and socially operative alarm calls doubtless contribute considerably to their protection against predators. Although bushbabies are dispersed during nocturnal activity, they remain in communication with one another by means of their calls (cf. p.167-91).

It is primarily the young bushbabies which are exposed to risk, and in two instances (one *Galago alleni*, one *Galago demidovii*) known young individuals disappeared completely from the study area. During the night, the mother carries her infant out of the nest and leaves it clinging to a thin branch. In the presence of any danger, the infant remains completely immobile and only drops from its support when the latter is shaken. At the instant when it releases its hold on the branch, the young bushbaby may utter a distress-call (if extremely frightened, or if it has actually been touched). It falls to the ground and 'freezes' after executing two or three hops. On a number of occasions, the author attempted to capture young Demidoff's bushbabies which had exhibited this escape response. Once they have 'frozen' on the ground, it is very difficult to spot them. If the infant has uttered distress-calls, the mother rapidly arrives on the scene and utters alarm calls in turn, sometimes approaching the 'aggressor' to a distance of less than one metre. As soon as the latter has moved away, the mother utters a faint, guttural call (r.r.r.r. ...), to which the infant responds with a summoning call. The distress-calls will evoke a response from any lactating female and even, to a lesser extent, from any other individual (male or female) in the locality. All will approach the infant in distress, frequently accompanying their advance with alarm calls. Jones (1969) observed in Rio Muni a *Galago demidovii* uttering alarm calls near to a conspecific which had been captured on the ground by a viper. The local village children in Gabon, who sometimes hunted *Galago demidovii* with a crossbow, made use of imitation distress-calls in order to attract their quarry. The author similarly used this technique of imitation in order to ensure identification of marked bushbabies at close quarters.

4

Activity Patterns and Sleeping Sites

A. ACTIVITY PATTERNS

The five prosimian species studied are all completely nocturnal in habit. Some authors have described *Galago demidovii* and/or *Arctocebus calabarensis* as diurnal or partially diurnal (Sanderson, 1940; Napier and Napier, 1967; Jones, 1969), but this is doubtless no more than a reflection of the fact that *Galago demidovii* sleeps in foliage and may flee in broad daylight when disturbed. Sanderson, at least, probably regarded this bushbaby species as diurnal for this reason. A different phenomenon may have led Napier and Napier, or Jones, to this interpretation. On the basis of observations conducted in captivity, they were perhaps drawn to the conclusion that *Galago demidovii* and *Arctocebus calabarensis* are partially diurnal or at least crepuscular. It is true that, in captivity, the activity rhythms of these prosimians may be considerably modified; but this modification depends, so it would seem, on the quantity of food provided. An animal which is deprived of insects or has not eaten insect food in sufficient quantity will frequently wake up one or two hours before nightfall to eat the fruit provided in its food-bowl. Conversely, if a great quantity of insects is fed to a captive specimen, it will consume far more than the normal quantity and will then become active only one to three hours after nightfall the next night. (This has been observed by the author with all five prosimian species in his captive colony.)

In the forest, no such dramatic shifts in the time of activity have been observed, and it would appear that such effects are limited to the artificial disturbance imposed in captivity. The ambient light intensity is the principal factor regulating the onset of activity in these prosimians in their natural environment. In order to obtain exact data on these rhythms, all five species were followed under free-ranging conditions and the times of onset and arrest of activity were noted by direct observation after the diurnal sleeping sites had been previously located. In the course of these observations, an accurate light-meter was used to measure the light intensity from the sky in the open field.

The prosimians do not set off in search of food as soon as they wake up. Approximately half-an-hour before night-fall, they start a long bout of self-grooming whilst stretching and yawning. It is only when the light intensity reaches a certain critical low level that they move out from their sleeping site (nest or patch of foliage). Very similar behaviour has been observed with the sportive lemur (*Lepilemur mustelinus*) and with the fork-crowned lemur (*Phaner furcifer*) in Madagascar (Charles-Dominique and Hladik, 1971; Pariente, 1974). The galagines begin to move around under crepuscular conditions, *Euoticus elegantulus* becoming active at light intensities between 300 and 100 lux, *Galago alleni* at 50 to 20 lux, and *Galago demidovii* at 150 to 20 lux. *Perodicticus potto* and *Arctocebus calabarensis* only become active under full night-time conditions (the late onset of activity in lorisines is probably yet another aspect of cryptic behaviour) (Fig. 53).

For all five prosimian species, arrest of activity occurs as soon as dawn commences, before the point at which the light-meter began to respond (light intensity less than 0.25 lux). Here, the two lorisine species are the first to *cease* activity. (Fig. 54).

The values obtained and plotted for the light intensities involved (Figs. 53 and 54) relate to measurements taken from the open sky. The foliage filters off a large proportion of the available light, and the onset of nightfall is so rapid at the equator (half an hour for the transition from 500 lux to 1 lux and only a quarter of an hour for the transition from 50 to 1 lux) that the bushbabies are only active for 5-10 minutes under crepuscular conditions prior to attainment of full night-time.

Galago alleni always spends the day sleeping in hollow trees, whose opening is usually directed towards the sky. Before emerging, this bushbaby exhibits extensive grooming

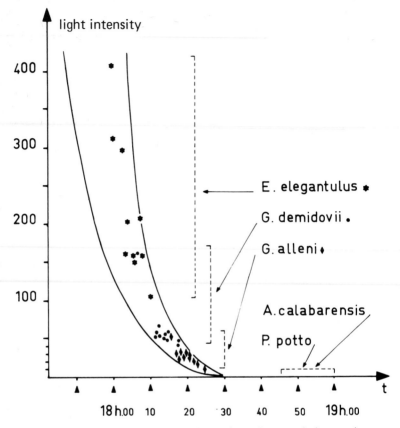

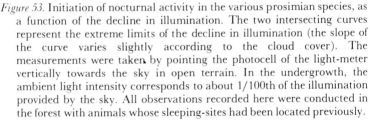

Figure 53. Initiation of nocturnal activity in the various prosimian species, as a function of the decline in illumination. The two intersecting curves represent the extreme limits of the decline in illumination (the slope of the curve varies slightly according to the cloud cover). The measurements were taken by pointing the photocell of the light-meter vertically towards the sky in open terrain. In the undergrowth, the ambient light intensity corresponds to about 1/100th of the illumination provided by the sky. All observations recorded here were conducted in the forest with animals whose sleeping-sites had been located previously.

behaviour interrupted by brief visits to the nest entrance. When the light intensity falls to 50-20 lux, the animal leaps out of its nest and will commonly head for trees in fruit.

One female Allen's bushbaby followed during the study made alternate use of two retreats situated 200 metres away from one another. The first nest was located in dense forest,

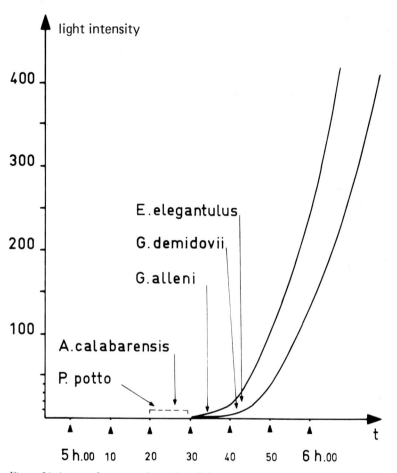

Figure 54. Arrest of nocturnal activity of the prosimian species as a function of light intensity, determined in the same way as for Fig.53. Return to the diurnal resting site occurs as soon as the first glimmer of dawn appears, but with the needle-clawed bushbaby and with Demidoff's bushbaby regrouping of social groups brings about slight delay in the complete arrest of activity.

whilst the second was in a tree close to the laboratory in an area where the established forest had been cleared leaving only an occasional tree here and there. With both retreats, the bushbaby made her exit at light intensities of 50-20 lux. In the dense forest zone, active locomotion occurred immediately after exit, whilst in the cleared zone the bushbaby waited until full nightfall had almost been achieved before entering into activity. In the meantime, the female remained concealed in foliage a dozen metres away from the nest exit.

It is obvious that it is the amount of light reaching the nest exit which determines the actual time at which the nest is left. In the course of the visits made to the exit, *Galago alleni* can assess the available light. In a clearing 25 lux corresponds to a light intensity in the undergrowth of less than 0.25 lux, and Allen's bushbaby will only move around actively when the actual ambient light intensity is below this value.

The other four prosimian species studied sleep out in patches of foliage during the daytime and they are hence able to follow the decline in light intensity in a continuous fashion. The long bout of self-grooming preceding the onset of activity, together with the correlation existing between the light intensity of the sky and the time of activity-onset, indicates that – as with *Galago alleni* – the ambient light conditions probably determine the commencement of activity.

During the course of the night, the five lorisid species rest from time to time, usually when they have full stomachs. These resting intervals have nothing in common with the diurnal sleeping phase; the animals do not move to their habitual sleeping sites and they do not actually sleep during such resting phases.

All five species continue to move around and to feed during light showers, but under conditions of heavy rainfall they commonly retreat into the nearest patch of dense foliage. The potto is the only one of the five species to remain active in heavy rainfall, continuing to feed even when the rainfall becomes really violent.

B. DIURNAL SLEEPING SITES

Most of the information provided in the literature about the diurnal sleeping sites of these prosimians is based on observations conducted in captivity. In the cages typically provided, the captive prosimians are not usually provided with patches of foliage comparable to those in which they would normally sleep. In most cases, they will retreat into any nest-boxes which are available, and this has led several authors to draw erroneous conclusions. It is for this reason that the potto, for example, is falsely reputed to sleep in hollow trees.

(i) *Perodicticus potto*

The potto nearly always spends the day sleeping on a relatively thin branch, hidden amongst foliage or in dense vegetation invaded by lianes (more than 100 observations of sleeping sites conducted in the forest). Pottos were sometimes seen sleeping at the tops of trees on branches exposed to the sun and the rain. During the hottest hours of the day, these animals stretch out without appearing to be particularly disturbed by the sun's rays, which are extremely intense in the equatorial zone. Under conditions of heavy rain, these same individuals remain rolled up in a ball on the branch. In such cases, only a superficial layer of the fur becomes wet; the rain runs off in rivulets without penetrating to the underlying wool hair, which remains completely dry. In most cases, however, this species selects sites which are relatively well sheltered under the cover of vegetation. Two pottos (2 males) followed by radio-tracking over a period of one month in primary forest failed to show any regular sleeping site within their home ranges. They simply moved into a patch of dense foliage close to the place where they happened to be when the sun began to rise.

(ii) *Arctocebus calabarensis*

As with the potto, the angwantibo typically sleeps in patches of foliage during the day, but always seeks out dense vegetation where it can sleep without exposure to the sun and the rain. Three angwantibos were observed during the study to die in less than an hour when accidentally exposed to the sun in a cage. In the rain, their fur is rapidly soaked, and on 10 occasions where sleeping sites were observed in the forest, individuals were seen to hide beneath dense leaf cover during tornadoes.

(iii) *Galago demidovii*

This species sleeps during the daytime either in specially-constructed leaf-nests or simply in suitably dense patches of vegetation (see also Vincent, 1968 and 1969). Usually, it is the females and juveniles which group together at such sleeping sites, whilst the adult males sleep alone (Charles-Dominique, 1972).

In the Congo, according to Vincent, the choice of the sleeping site (nest or simple branch-fork with leaf cover) depends upon the ambient temperature, with Demidoff's bushbabies only grouping together in nests during the coldest weather. In Gabon, on the other hand, it seemed that this choice was related to the presence or absence of infants. In nests where several adults (females) were present, it was almost always noted that there were also one or more infants or juveniles in the weight-range of 7-35 g. Under captive conditions, nest-building behaviour is elicited during periods of copulation and, particularly, during times when young are being reared. In Gabon, the birth maximum (see p.142) is found during the rainy season (hot weather period) and the minimum occurs during the dry season (cold weather period, with no rainfall at all). The construction of a nest would thus seem to be associated with protection against the rain, since the young are only covered with a sparse layer of hair at the time of birth.

The sleeping sites observed were always located in relatively dense vegetation in the shade (25 observations in the forest). Vincent (1968) points out that the selection of the *height* of the nest above the ground depends upon the height of the available vegetation cover, and this same conclusion was drawn from the author's observations conducted in Gabon. The same sleeping site (nest or entwined patch of lianes and branches) may be utilised on several successive days (see also Struhsaker, 1970); but the site is changed at once if the animals are disturbed, and such change may also occur spontaneously. Nevertheless, there are some zones in the forest which are particularly suitable and this explains observations of nests of different ages located in close proximity (Vincent, 1969).

(iv) *Galago alleni*

This is the only one of the five species studied which passes the day sleeping in tree-hollows. The 53 sleeping sites observed in the forest were all composed of hollow trunks with an opening above, forming a kind of chimney, or opening to one side. The internal diameter of such 'chimneys' is very variable, ranging from 10 to 80 cm. In some cases, the animal rests at the base of the hollow and may carry in a few leaves as lining material. In other cases, the bushbaby may pass the entire day clinging to the inside face of the chimney (3 observations, including a female accompanied by a juvenile 1-2 months old). There does not seem to be any competition involved in the search for suitable tree-hollows, since experience shows that in the forest it is necessary to examine a large number before finding one which is occupied. On the other hand, one large 'chimney' (a dead trunk which had remained standing) was occupied by 5 or 6 bats, a gliding rodent (*Anomalurus fraseri*) and one Allen's bushbaby, all clinging to the inside of this one retreat. On a separate occasion, a female flying squirrel (*Anomalurus fraseri*) and her offspring were observed clinging at the entrance of a relatively small tree-hollow which had been occupied by two female *Galago alleni* a short while previously. The next day, these same two female Allen's bushbabies (which were being followed by radio-tracking) selected a different tree-hollow as a retreat.

Within its home range, each individual *Galago alleni* has available several retreats which are used in succession. One female followed by radio-tracking was found to occupy 12 different sleeping sites in the course of 32 separate observations, with certain sites being used more frequently than others. Another female was seen to use 7 different sleeping sites in the course of 16 observations, while a third used 7 sites during 26 observations. These sleeping sites seemed to be scattered evenly throughout the home range. Males, by contrast, sleep individually in one of a cluster of sleeping sites localised in a small area of the home range. Only on rare occasions do they use a sleeping site normally occupied by a female (2 cases out of 28 observations conducted by radio-tracking). As with the angwantibo, Allen's bushbaby has fur which is rapidly soaked in the rain, and this is doubtless correlated with the typical use of tree-hollows by this species. Although such tree-hollows are commonly occupied by single individuals, two or three adult

5

Population Parameters

A. POPULATION DENSITY AND DISTRIBUTION

From the very outset of the study, the attempt was made to estimate population densities for each of the prosimian species concerned. Details of the techniques of assessment are provided in a previous publication (Charles-Dominique, 1972), and only the basic principle will be summarised here.

Counts were made of all animals seen at night with a headlamp along measured transects. For each animal sighted, the distance from the transect axis was estimated, thus permitting the establishment of a 'visibility profile' for each type of environment (undergrowth; primary forest canopy; secondary forest; etc.). On the basis of this 'visibility profile' it was possible to determine the width of the strip of forest effectively penetrated by the headlamp. Only animals located within this strip were taken into account for calculation of average population densities per square kilometre (utilising data derived from 580 km of transect covered, corresponding to 148 km of different transects or pathways explored once or several times each). Over and above this, the raw data have been refined with information derived from systematic trapping over large areas, detailed study of individual home ranges and intensive investigation of social behaviour through radio-tracking. Taking all of these lines of investigation into account, the following population densities have been estimated for primary forest:

Perodicticus potto	8-10 per square kilometre
Arctocebus calabarensis	2[1] per square kilometre
Galago demidovii	50-80 per square kilometre
Galago alleni	15-20 per square kilometre
Euoticus elegantulus	15-20 per square kilometre

These population densities for the five prosimian species are low compared with those calculated by Bearder and Doyle (1974) for *Galago senegalensis* (87-500 per square kilometre) and *Galago crassicaudatus* (72-125 per square kilometre) in wooded savannah and gallery forest of South Africa. It should be noted, however, that according to these two authors the lowest population densities (103 and 72, respectively) were found in an area where the two bushbaby species are sympatric. In addition, this greater concentration of lorisid species in dry zones is related to the fact that a smaller number of species is found to coexist in such areas. The same situation is found in Madagascar, where the population densities of nocturnal lemurs (determined by the same techniques) are generally very high: *Lepilemur mustelinus* = 200-450 per square kilometre (Charles-Dominique and Hladik, 1971); *Microcebus murinus* = 250-360 per square kilometre (Charles-Dominique and Hladik, 1971), *Cheirogaleus medius* = approx. 250 per square kilometre (unpublished data). In this case, these high values are correlated with the poverty of the Malagasy mammalian fauna in terms of numbers of different mammal species. There are only 15-30 sympatric mammal species (according to the forest type), as opposed to 120 mammal species in the rain-forest of Gabon. It should also be noted that Walker (1974) has pointed out that the proportions between fossil lorisines and galagines in the East African Miocene are calculated to be similar to those found in Gabon.

The primary forest, which is markedly heterogeneous in structure, presents a mosaic of biotopes of different kinds, depending upon the hydrographic conditions, the distribution of plant species and the occurrence of tree-fall zones in which all stages of forest regeneration are to be found. Some of these

---

[1] The average figure of 2/sq.km relates to primary forest taken as a whole. However, angwantibo populations are concentrated in well-defined biotopes (see p.131), where the average density actually attains an average value of 7/sq.km.

biotopes seem to be more suitable than others for occupation by prosimian species, in that repeated night-time population counts reveal concentrations of animals in certain zones which lie across the transects. For example, forest which is flooded from time to time seems to be particularly rich in pottos (28 per square kilometre, as opposed to the average figure of 8-10) and less rich in needle-clawed bushbabies (3 per square kilometre, as compared to the average of 15-20). *Galago demidovii* is particularly attracted by zones adjacent to roads, where suitable vegetation is better developed (117 per square kilometre, compared to the average density of 50-80). These population 'nuclei' are the outcome of species-specific ecological preferences. When the home range of a potto includes a section of forest which is exposed to flooding, this area is utilised more frequently than other, drier, parts of the range. Conversely, *Euoticus elegantulus* – which principally seeks out gums formed by trees of the family Mimosaceae – pays fewer visits to flooding zones, which are less favourable for the growth of such tree species. Fig.55 illustrates the correlation existing between the frequencies of sighting of *Euoticus elegantulus* and the distribution of *Entada gigas*, a liane species which yields 80% of the gums exploited by the needle-clawed bushbaby on the Ipassa plateau. It should be added that the patterns of social organisation of the five lorisid species (see p.160) necessitate juxtaposition and some degree of overlap of the home ranges, which tends to reinforce the densities of individuals in zones providing the greatest expanse of favourable biotope. Thus, the so-called 'population nuclei' which may be identified represent only a relative degree of isolation between local areas of abundance.

The case of the angwantibo provides an extreme example. In primary forest, this species is confined to the undergrowth in old tree-fall zones where the regenerating forest is particularly rich in young lianes. Fig.56 shows the distribution of such biotopes on the Ipassa plateau. Systematic searches for fallen tree-trunks showed that they had generally been uprooted or shattered by natural forces and that most of them had fallen towards the south-east. Tornadoes follow certain common pathways which are readily recognised by the local villagers as 'rainfall strips' along which the trees are particularly exposed to damage. The area of the forest studied

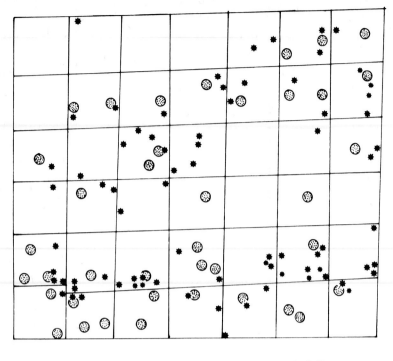

Figure 55. Diagram illustrating the distribution of a gum-producing liane (*Entada gigas*) and sightings of needle-clawed bushbabies during the night. (Circles = lianes; stars = bushbaby sightings.) The data were obtained by equally distributed surveys conducted along transects at intervals of 100 m. (77 sightings of *Euoticus elegantulus* in 42 1-hectare squares containing 39 identified bases of *Entada gigas*). In this region, the needle-clawed bushbabies feed primarily on gums from *Entada gigas* and follow the heterogeneous distribution of this liane species. (N.B. The transects, which were not exactly parallel, are indicated by lines.)

was in fact located in one such strip on the plateau, and the pattern of distribution of tree-fall zones follows various corridors with a south-easterly orientation, running together at various points and eventually leading to a point on the promontory overhanging the river (Fig.57).

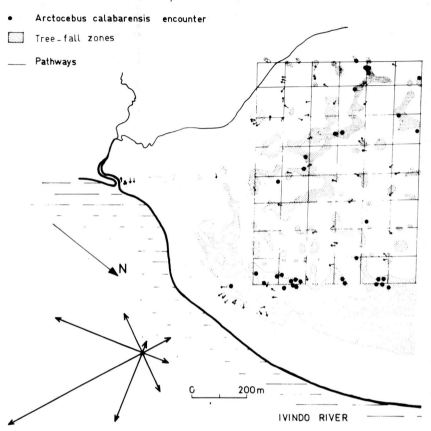

Figure 56. Disturbed forest areas in the primary forest of Ipassa (dotted zones). These areas correspond to repeated tree-falling which permits the development of a low-profile, luxuriant vegetation rich in lianes. The direction of recent tree-falling is indicated by the small arrows. Most of these lie in the quadrant East South-East (diagram), which corresponds to the principal direction taken by the tornadoes. Almost all sightings of angwantibos (circles) have been made in disturbed forest areas, whose distribution depends upon topographical features.

The distribution of angwantibos in the forest exactly coincides with the occurrence of tree-fall zones. In other areas of the forest which are much less exposed to tornadoes and where tree-falling is less common, no angwantibos were ever sighted.

Figure 57. Photographs of disturbed areas in primary forest (Photo: A.R. Devez)

B. LONGEVITY AND POPULATION REPLACEMENT

There is very little information available on the longevity of lorisid species in captivity.

(i) *Perodicticus potto*

Jarvis and Morris (1960) record the survival of one individual in captivity for a period of 9 years. In the author's colony a male and female captured as adults 7 years ago are still continuing to breed. In fact, the development of the teats indicated that the female was multiparous when captured, and she must now be at least 9 years old. In the field, one female captured as a multiparous adult in 1966 was still alive seven years later in 1974, in the same area of forest, and must at that time have been nine years old or more. However, the field-study covered far too few individual pottos to provide exact information on the average life-span under natural conditions. Most of the individuals examined for the first time were already adult, so the following data provide only a minimum assessment for population replacement. Continuous presence of 6 individuals identified for the first time as adults in the study area was noted as follows:

adult males	1 year; 1 year; 3 years
adult females	3 years; 3 years; at least 7 years

(ii) *Arctocebus calabarensis*

Longevity in captivity for this species is reported as 5 years (Jacobi – cited by Vincent, 1969) and 4 years (Jarvis and Morris, 1960). No information is available from the field.

(iii) *Galago demidovii*

Data now available indicate that this species survives longer than was suggested in the literature by Cansdale (1960: 4 years) and by Vincent (1969: 3 years). In the author's colony, of 8 individuals captured as adults 4 years ago and one captured 5 years ago (5 males; 4 females), only 3 have so far died, all from accidental causes. One female died whilst giving birth, a male died following a wasp-sting and a second male died following a long period spent in a cage where it was subjected to continuing aggression from a conspecific. The 6 surviving individuals are in good condition and are still breeding. In view of the fact that sexual maturity is first attained at about the age of one year and that the females concerned were already multiparous when captured, the *minimum* ages of the three surviving males must be 6 years, 5 years and 5 years and all three surviving females must be at least 6 years old. Exact data on the potential longevity of this species will only be available several years hence when animals born in captivity have been observed for a sufficient period of time.

In the field, extensive investigation conducted over five successive years on a natural population of about 40 individuals (see p.150) yielded data for estimation of replacement in this population. In fact, purely on the basis of the presence or absence of marked individuals it is impossible to state with certainty whether bushbabies which 'disappear' have died, since it is not known whether all marked individuals will eventually be re-trapped (if still in the vicinity). Nevertheless, it is possible to produce a diagram (Fig.58) providing some indication of replacement within the population and – for the females, which are always sedentary – a minimum figure for the real longevity characteristic under natural conditions. It can be noted that the males disappeared from the study area much more rapidly than females during the first year of investigation. Thereafter, the curve flattened off and became broadly parallel to that obtained for the females (Fig.58). This marked drop in the first year reflected the fact that many males first identified during the 'vagabond stage' (see later) move away from the study area later on and thus avoid the traps which are set. As far as the females are concerned, half of the population was renewed in the course of three years, and after five years only 2 of the original 16 females remained. The population must therefore be completely replaced approximately every six years, if these data accurately reflect the state of affairs existing under natural conditions.

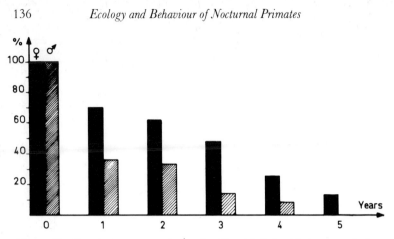

Figure 58. Population renewal in the Demidoff bushbaby population
followed by trapping and marking in secondary forest, after 1 year (N =
26; N' = 36), after 2 years (N = 16; N' = 15), after 3 years (N = 32; N' =
43), after 4 years (N = 32; N' = 43), after 5 years (N = 16; N' = 26). N =
number of females on which the average is based. N' = total number of
males on which the average is based. The vertical bars indicate the
percentage of the original animals in each year's sample.

(iv) *Galago alleni*

Two females of this species first examined in a multiparous
condition were recaptured in the forest 6 years later, when
they must both have been at least 8 years old. With a natural
population studied in a primary forest area, almost all of the
females captured were recaptured 1 year later. On the other
hand, the 2 adult males previously captured in the area had
been replaced by 2 newcomers after a year. It therefore seems
likely that the males engage in intense competition (cf. the
data on sex-ratios on p.138 and the discussion of social
behaviour on p.162).

(v) *Euoticus elegantulus*

No information is available in the literature for this species,
and the author's captive stock has been established too
recently to provide data on longevity at the present time.

C. SEX-RATIOS

Determination of the sex-ratio under natural conditions for any of the prosimian species requires collection of a uniform sample of a population. The method of capture is therefore very important, for males and females quite often behave in different ways. In general, males are more active than females and have larger home ranges. Trapping with a noose or with baited traps, for example, reflects such behavioural differences with *Galago demidovii*. Martin (1972) has observed similar differences in the apparent sex-ratio determined for *Microcebus murinus* (lesser mouse lemurs), according to the method of study (trapping at night; examination of nest occupants; direct observation of the sexes of sighted animals at night). As an example of the different values for the sex-ratio obtained with different methods of capture, one can take the following data obtained for *Galago demidovii* in Gabon:

Capture with a noose-trap following chance encounters at night (N=37): males = 67%; females = 33%

Capture with baited traps (N=92): males = 60%; females = 40%

Capture in nests (N=11): males = 18%; females = 82%

Hunting with a rifle (N=39): males = 49%; females = 51%

These figures demonstrate quite clearly that males move around in the forest more than females (leading to high figures for capture with nooses), and this same tendency leads them to discover more rapidly than the females any traps baited with banana.

Capture of animals occupying nests is also inaccurate as a means of establishing the sex-ratio, since the individuals found in nests are primarily adult females, their offspring and (sometimes) single dominant males. Hence, hunting with a rifle at night would seem to provide the most reliable data, since the animals are collected by an observer moving around during the animals' nocturnal activity period (see Table 5).

These data, which are insufficient for precise calculation of the sex-ratio, do permit us to draw two overall conclusions:

Table 5 Observed sex-ratios of adults and juveniles of the five prosimian species in Makokou

	% ♂ ad.	% ♀ ad.	N ♂/N ♀	% ♂ juv.	% ♀ juv.	N ♂/N ♀
Perodicticus potto	= 35%	= 65%	15/27	= 50%	= 50%	12/11
Arctocebus calabarensis	= 50%	= 50%	20/19	= 50%	= 50%	13/14
*Galago demidovii**	= 50%	= 50%	19/20	= 50%	= 50%	21/23
Galago alleni	= 20%	= 80%	4/17	= 50%	= 50%	6/5
Euoticus elegantulus	= 55%	= 45%	28/22	= 50%	= 50%	3/3

*Vincent (1969) obtained the following percentages for adult males trapped in 2 areas of the Congo: 53.64% and 56.61%. On the other hand, a figure of 59.57% was found for males at birth.

1. In all five species, at birth the numbers of males and females would appear to be roughly equivalent.
2. With *Perodicticus potto* and *Galago alleni*, the percentage of males is diminished among the adults. It will be seen from the descriptions of social behaviour (p.215) that the males of these two species are in fact the most aggressive in their encounters with conspecific males.

D. BREEDING SEASONS

Galago senegalensis is the best known of the lorisids from the point of view of seasonality of breeding (Butler, 1957, 1960, 1967; Haddow and Ellice, 1964; Doyle, Andersson and Bearder, 1971). In the arid zones occupied by this bushbaby species, breeding occurs during the 'best season' from the point of view of abundance of food: December to April in the Sudan (Butler, 1967); October to the beginning of November and the end of January to the beginning of February in South Africa (Doyle et al., 1971). There are two birth-peaks separated by four months in South Africa, corresponding to two successive gestation periods for each female (Doyle et al., 1971). Thus, the reproductive potential of this species is quite high, since each female breeds twice a year and since more than half of the females give birth to twins or even (occasionally) triplets.

With the five Gabonese lorisid species, the situation seems to be quite different on the basis of the much more limited information available on their reproduction under natural conditions. (For further data on these species, see also Jewell and Oates (1969) with respect to the breeding of lorisids in Biafra, and Vincent (1969) for information on the reproduction of *Galago demidovii* in Congo Brazzaville.)

In order to permit interpretation of the data on seasonality of breeding, gestation periods are given for all five Gabon prosimian species in Table 6.

(i) *Galago demidovii*

According to Vincent (1969) in the Congo Brazzaville region this species gives birth predominantly in the periods 15th September to 26th October and 16th January to 12th February. Over the rest of the year, scattered births were also

Table 6. Gestation periods and developmental characteristics of the five Gabon lorisid species. Figures for *Galago senegalensis* are included for comparison.

	Gestation period (days)	Weight at birth (gm)	Age at Weaning (days)	Eruption of 3rd molar (days)	Attainment of adult weight (months)	Age at sexual maturity (months)
Galago demidovii 60 – 80g	111-112-114[1]	5-10[1] 5-12[5]	40-50[1]	60-65[1]	5-6[1,5]	8-10[1] 6-10[5]
Galago alleni 250 – 300g	133[1]	24[1]	?	100[1]	6-8[1]	8-10[1]
Euoticus elegantulus 300g	?	?	?	≏ 100[1]	≏ 10[1]	?
Galago senegalensis 150-200g	120[3,8] 122-125[4] 139 ± 3[7] 144-146[2]	8.5-15.5[6] 15.6-22.6[7]	60-80[4]	?	?	7[6]
Perodicticus potto 1 kg	193[1]	52[1] 30-42[9]	120-180[1]	≏ 180[1]	8-14[1]	≏ 18[1]
Arctocebus calabarensis 250g	131-136[10]	24-30[1]	≏ 100-130[1]	≏ 100[1]	8-9[1]	≏ 9-10[1]

[1] Charles-Dominique (pers.obs.)
[2] Manley, 1966
[3] Lowther, 1940
[4] Doyle et al, 1967
[5] Vincent, 1969
[6] Doyle et al., 1971
[7] Cooper (in Doyle et al., 1967)
[8] Sauer and Sauer, 1963
[9] Cowgill, 1969
[10] Manley, 1967

reported. In Gabon, however, Demidoff's bushbaby appears to exhibit a seasonal pattern of reproduction different from that reported for the Congo. In the Gabon study area, 47 births were recorded in the forest between 1965 and 1973. For purposes of comparison between different months a monthly percentage was calculated for the number of recorded births with respect to the number of adult females examined. Births occur virtually throughout the year (Fig.59), but there is a maximum in the period January-April inclusive (the period of greatest abundance of fruit and insects) and a minimum in June/July (when insects and fruits are scarce – see Fig.10, p.19). In captivity, births also occur throughout the year, both under conditions of cyclically increasing and decreasing daylength (natural daylight variation of the Paris region) and under constant light conditions (11 hours of darkness: 13 hours of light, as in Gabon). In captivity, fertile mating has also been observed a few days after parturition on four separate occasions, reminiscent of the post-partum oestrus recorded for *Galago senegalensis* by Butler (1960).

Under natural conditions, it would seem that *Galago demidovii* females breed only once a year. Of 85 females examined by dissection at different times throughout the year, 11 were found to be lactating (viz. 12%), whilst 31 were gestating (viz. 35%). Lactation lasts 45 days (i.e. 12% of the year), whilst the gestation period is 111-114 days (i.e. 33% of the year). The correspondence between these two sets of figures indicates that there is approximately one parturition per year per female, on average. It is unlikely that in the wild females mate immediately after birth, as may happen in captivity, since none of the 31 gestating females examined at various times of the year was lactating. Examination of 18 pregnant females captured in the forest and subsequently dissected showed that 14 had a single embryo, whilst only 4 had two embryos. Hence, the reproductive potential is considerably less than that of *Galago senegalensis*; but low fecundity has been found to be characteristic of a large number of animals in the equatorial rain-forest block (Brosset, 1971).

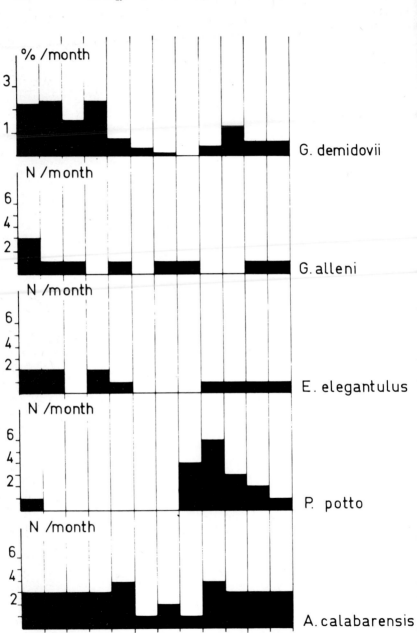

(ii) *Galago alleni*

Little information is available on the reproduction of this species, but the pattern would seem to be generally similar to that found with Demidoff's bushbaby in Gabon. The 10 births recorded in the field were spread more or less over the entire year, perhaps with a peak in January (Fig.59). Among the 15 adult females captured in the forest and subsequently dissected, 3 were lactating, as compared to 10 non-lactating (state of teats not checked in 2 females), whilst 4 were gestating, as opposed to 11 non-gestating. No female captured in the forest was found to be lactating and gestating at the same time. With the 4 pregnant females trapped in the forest and dissected, one embryo was found in 3 cases, and twin embryos were found in the fourth.

(iii) *Euoticus elegantulus*

With this species, 11 births were recorded under natural conditions; their distribution over the year seemed to follow the pattern found in the other two galagines. Among the 17 females captured and dissected at various times of the year, 9 were gestating as opposed to 8 non-gestating, and 6 were lactating as opposed to 7 non-lactating (state of teats not checked in 4 females). In two cases, a lactating female was found to be carrying an early embryo (weights: 0.4 gm and 3 gm, respectively). All of the 9 pregnant females dissected had a single embryo.

(iv) *Perodicticus potto*

The 17 births recorded for this species over the period 1965-1973 were all found to occur in the months August-January inclusive. Weaning hence occurs principally in January-

Figure 59 (left). Annual distribution of births in the 5 lorisid species. Results are based on data obtained in the period 1965-1973, covering the 12 months of the year virtually uniformly when pooled. For *Galago demidovii*, where large numbers of observations were available, the percentage of births per month recorded under natural conditions (with respect to the total number of adult females examined) is given. With the other species, for which fewer data are available, the total numbers of births recorded under natural conditions for each month are given.

March, which is the time when fruits are most abundant (see Fig. 10,'p.19). The data for the ratio between lactating and non-lactating females (9:9) and for the ratio between gestating and non-gestating females (4:14) collected throughout the year appear to be contradictory. It is possible that gestating females are more discreet in the forest than lactating females, which may need to ingest a large quantity of food and thus expose themselves to greater risk. One female followed in the forest was observed to give birth on four successive occasions: October 1966, November 1967, end of November 1968, 20th November 1969. Another female in the forest provided the following records: birth at the end of January 1969, birth at the end of October 1969, (no observations conducted in 1970), gestating in August 1971, gestating in July 1972, gestating in July 1973. (N.B. the gestation period is approximately six months – see Table 6.) It is thus apparent that female pottos usually breed once a year. Normally, there is one infant in each litter, but one case of twins has been recorded in captivity (Cowgill, 1974).

(v) *Arctocebus calabarensis*

For angwantibos, 33 births recorded in the field were spread more or less throughout the year with an apparent minimum in the period June-August inclusive. Contrary to the situation found with the potto, female angwantibos can breed more than once in a year. With 13 females dissected at various times of the year, 10 were found to be gestating, as opposed to 3 non-gestating, and 5 were found to be lactating, as opposed to 3 non-lactating (state of the teats not checked in 5 females). Four of the 5 lactating females were found to be carrying an embryo. (There is typically one infant per female.) In three cases, it was possible to confirm that conception had followed the previous birth by a matter of a few days.

  In captivity, several lorisid species (*Perodicticus potto, Loris tardigradus, Galago senegalensis, Galago crassicaudatus, Galago alleni, Galago demidovii*) have exhibited seasonal sexual activity, with the females showing a polyoestrous pattern over part of the year (Petter-Rousseaux, 1962; Ramaswami and Anad Kumar, 1962, 1965; Doyle and Bekker, 1967; Doyle et al., 1967; Charles-Dominique, 1968 and unpublished data;

Vincent, 1969). Yet some of these species maintained under different conditions in captivity (*Perodicticus potto, Arctocebus calabarensis, Galago senegalensis, Galago crassicaudatus, Galago alleni*) have exhibited regular sexual activity occurring throughout the year (Buettner-Janusch, 1964; Ioannou, 1966; Manley, 1966; Charles-Dominique, 1968 and unpublished data).

Data obtained in captivity are often contradictory to information obtained in the field. Whereas with *Galago senegalensis* and *Perodicticus potto* births are restricted to a particular period of the year under natural conditions, with the other African lorisids births occur throughout the year, usually showing no more than a seasonal peak in frequency. Nevertheless, the field data obtained in Gabon indicate that with the potto and at least two of the bushbaby species each female generally breeds only once a year, and this implies relatively long periods of anoestrus. It may be assumed that with species exhibiting seasonal reproduction certain factors have brought about synchronisation of oestrus, whilst with the other species such synchronisation is only partially determined.

With the Malagasy lemurs, where all species studied give birth during the austral spring and summer, annual variation in daylength (photoperiod) determines reproductive periodicity (Petter-Rousseaux, 1962, 1968, 1974). This would not seem to be the case with the lorisids, which generally live in tropical and equatorial areas where the climate is essentially dependent on the rainfall regime, varying from one region to another independently of daylength. The question of the timing mechanism for seasonal reproduction in these regions remains obscure, both for the lorisid species and for numerous other equatorial vertebrate species whose reproductive cycles are related to periods of high rainfall (Dubost, 1968; Jewell and Oates, 1969).

6

Social Behaviour

If one considers the number of publications dealing with the social life of the primates, it rapidly becomes apparent that the Lorisidae, along with the other nocturnal primates, have generally attracted little attention from naturalists. The latter have usually preferred to devote themselves to the study of 'more social' diurnal primate species, often phylogenetically closer to man and in any case easier to observe in the field. Nocturnal, 'solitary' prosimian species are of special interest, however, in that they to some extent represent a primitive evolutionary stage whose study may yield vital information for the understanding of the phylogenetic history of the Order Primates.

As a result of this relative neglect of the nocturnal prosimians in studies of primate social behaviour, it is necessary first of all to dispose of a number of misunderstandings which have arisen in the literature. 'Solitary' is not the opposite of 'social' but of 'gregarious'. A number of authors (e.g. Crook and Gartlan, 1966) have contrasted 'social' primates with 'solitary' primates, as if solitary life in some way excludes the possibility of social behaviour. In fact, although they are all solitary when active at night, the lorisids communicate with one another by means of vocalisations and also through olfactory signals which usually pass completely unnoticed by the human observer. It is an anthropomorphism to consider as 'more social' gregarious primates whose communicative behaviour is more obvious to our senses. It will be seen from the following account that the activity of the five lorisid species is in fact closely dependent upon the social environment (relative

Table 7 Comparative figures for nocturnal sightings of the five lorisid species, showing percentages of animals encountered singly or in groups of various sizes

	Isolated %	2 together %	3 together %	4 together %	5 together %	Mother & infant %
Perodicticus potto N = 105	96%	2%				2%
Arctocebus calabarensis N = 99	97%	1%				2%
Galago demidovii N = 263	75%	21%	2%	0.5%	0.5%	1%
Galago alleni N = 97	86%	8%	4%			2%
Euoticus elegantulus N = 103	76%	17%	2%	1%		4%

positions of home ranges; signals; visiting behaviour; defence; etc.).

It is, of course, true that during their nocturnal activity periods the lorisids are almost always encountered as single individuals, as is evident from figures obtained for nocturnal sightings within a single light-beam in the forest during the present study (Table 7).

The two lorisine species, which move around slowly in the forest as described above (p.69), do not usually meet up together, as is sometimes the case with the three bushbaby species, whose saltatory mode of locomotion permits greater mobility within the home range. Nevertheless, even the bushbabies, when seen at night, are solitary most of the time, and if two or three are seen together they are generally separated from one another by a distance of 10-30 metres. Accordingly, study of the social behaviour of these primates in captivity necessitates the use of methods different from those utilised for investigating diurnal monkey and ape behaviour. For example, studies of dominance and rank-order with respect to food within a group artificially assembled in a laboratory cage has little relevance to the situation existing under natural conditions.

The lorisines always sleep singly during the daytime, with the exception of mothers which collect their 'parked' infants before daybreak. The galagines, on the other hand, commonly form small sleeping-groups (see discussion of mother-infant relationships on p.227). This alternation between solitary life (active phase) and gregarious life (resting phase) among galagines had been noted by several authors at various times in the past (Sanderson, 1940; Malbrant and Maclatchy, 1949; Sauer and Sauer, 1963; Haddow and Ellice, 1964; Vincent, 1969), but generally no details were given for the ages and sexes of the animals in the groups. Such groups were, in fact, typically composed of adult females and their offspring in the three species studied in Gabon. The adult males generally sleep separately. These sleeping-groups will be considered in more detail when detailed social relationships are discussed (p.241).[1]

[1] Petter (1962) noted similar sleeping-groups in the lesser mouse lemur (*Microcebus murinus*), and it has now been shown (Martin, 1972a) that such groups are usually composed predominantly of adult females and their offspring.

The first concern of the field-study was surveillance of marked animals moving freely in the forest. In order to achieve this, a trapping system was developed by carefully and progressively habituating the naturally cautious prosimians to approach and eventually to enter inoperative traps. In the forest, all of the prosimians rapidly come to feed at baskets of lianes which are regularly provisioned with bananas. A trap was progressively brought closer and closer to such a liane basket, and then regularly baited with bananas, first placed on the trap and then, eventually, inside. After a period of 10-15 days, the prosimians would enter such a trap unhesitatingly and it was then possible to set the trap-door mechanism. When this technique is employed, the prosimians do not usually make an association between the trap and the event of capture, but between the *site* involved and the event of capture. Thus, by moving the traps around, it is possible to recapture prosimians on subsequent occasions. Of course, certain individuals remain very difficult or even impossible to trap. Further, the long dry season – in which food is very scarce in the forest – is the most suitable period for such trapping. In fact, *Euoticus elegantulus* and *Arctocebus calabarensis* are non-frugivorous (see Fig.17) and it proved impossible to capture any individuals of these two species at any time of the year. Thus, data on home ranges was only obtained for the other three prosimian species. In the period 1968-1973, the trapping programme yielded 282 captures of *Galago demidovii* (76 different individuals), 90 captures of *Galago alleni* (18 different individuals), and 70 captures of *Perodicticus potto* (20 different individuals) – giving a grand total of 442 captures. All the animals captured were marked with notches on their ears arranged in a variety of combinations. In order to permit their subsequent recognition at a distance, they were also marked in a number of other ways: clipping of different zones of the tail fur (*Galago* species), discoloration of the pelage (*Perodicticus potto*), attachment of coloured collars (*Galago alleni* and *Perodicticus potto*), and – in the course of the study visit in 1973 – radio-tracking (8 *Galago alleni* and 2 *Perodicticus potto*). Each time that a marked animal was recaptured, its body-weight and sexual condition were noted.

A. ORGANISATION OF HOME RANGES

With the exception of young males at puberty, which pass through a nomadic ('Vagabond') stage, all of the lorisids studied were found to be sedentary in habit. Pottos and Demidoff's bushbabies were studied in the same areas over four and five successive years (respectively), and there were many cases of individuals being recaptured at intervals of two or three years. On the basis of such long-term records, diagrams of home range organisation can be drawn up. Intensive study of *Galago demidovii* in secondary forest in 1968, involving successive captures (137 captures in all for 22 sedentary individuals and 8 vagabond males), provided a fairly detailed picture of the home range relationships for both males and females[1] (Figs. 60, 61, 62 and 63). In June/July 1973, a combination of trapping and (particularly) radio-tracking permitted definition of the home-range boundaries of a population of *Galago alleni* in primary forest, on the basis of 385 recordings of location (Fig.64). Finally, data obtained for a population of *Perodicticus potto* in the period 1967-1973 in secondary forest provided an outline of home range relationships in this species (see Figs. 65 and 66). In all cases, home range limits were established on the basis of trapping positions for each individual, direct observation of the movements of marked animals and radio-tracking information.

At first sight, there is a great degree of similarity between all three species in the arrangement of the home ranges. In every case, overlap of female home ranges is very variable, whilst individual home ranges of males scarcely overlap one another although they overlap one or more female home ranges. This arrangement has also been found with the lesser mouse lemur, *Microcebus murinus* (Martin, 1972a), with the sportive lemur, *Lepilemur mustelinus* (Charles-Dominique and Hladik, 1971), with the Senegal bushbaby, *Galago senegalensis* (Bearder and Doyle, 1974), with the Bornean tarsier, *Tarsius bancanus* (Fogden, 1974) and with numerous other nocturnal mammals (e.g. the African palm civet *Nandinia binotata*, Charles-Dominique, in prep.; aquatic chevrotain, Dubost, in prep.;

---

[1] For details of the grid system used in systematic trapping, see Charles-Dominique, 1972.

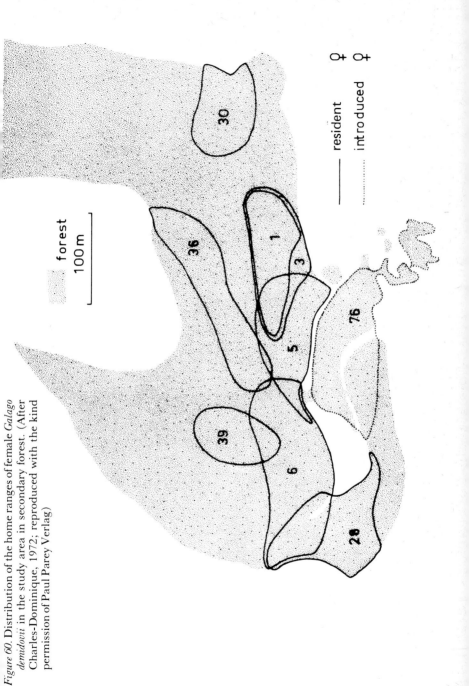

Figure 60. Distribution of the home ranges of female *Galago demidovii* in the study area in secondary forest. (After Charles-Dominique, 1972; reproduced with the kind permission of Paul Parey Verlag)

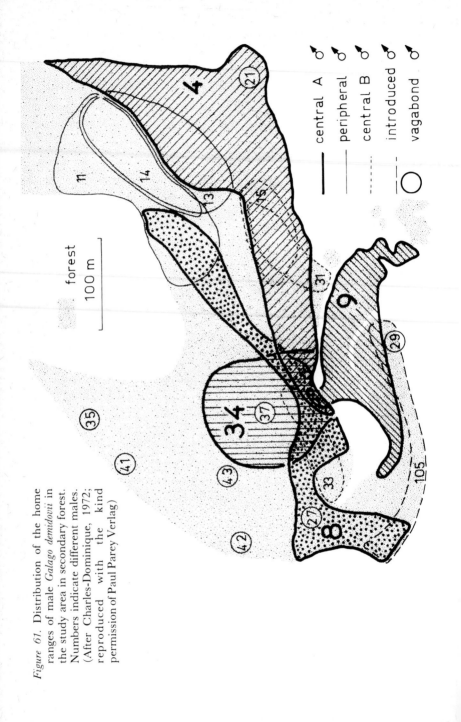

Figure 61. Distribution of the home ranges of male *Galago demidovii* in the study area in secondary forest. Numbers indicate different males. (After Charles-Dominique, 1972; reproduced with the kind permission of Paul Parey Verlag)

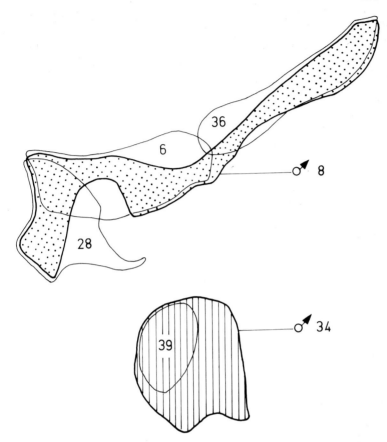

Figure 62. Overlapping of female home ranges by the home range of a polygamous adult male (♂ 8) and by that of a monogamous adult male (♂ 34) in the population of *Galago demidovii* investigated in secondary forest. (See Figs. 60 and 61 for details)

tenrecs, Eisenberg and Gould, 1970; sloth, Gregory, 1973). It would appear that this basic pattern may have been a primitive feature of the placental mammals.

On the other hand, home range *size* varies considerably from species to species, and there are also marked intraspecific differences according to sex and (among males) social status, as shown in Table 8.

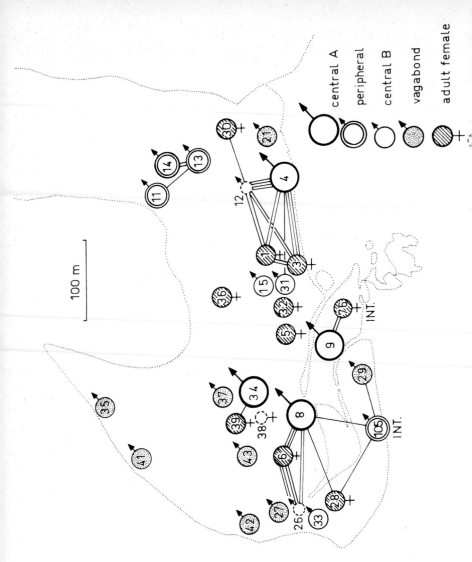

Figure 63. Social organisation of the population of *Galago demidovii* in secondary forest. Each individual is represented at the centre of its home range by its reference number and by a symbol corresponding to its social category. The lines indicate simultaneous capture of animals in one trap. One can distinguish four groups, each of which is centred around a Central A male, whilst a fifth group is constituted by peripheral, bachelor males. (INT = strange animals introduced experimentally). (After Charles-Dominique, 1972; reproduced with the kind permission of Paul Parey Verlag.)

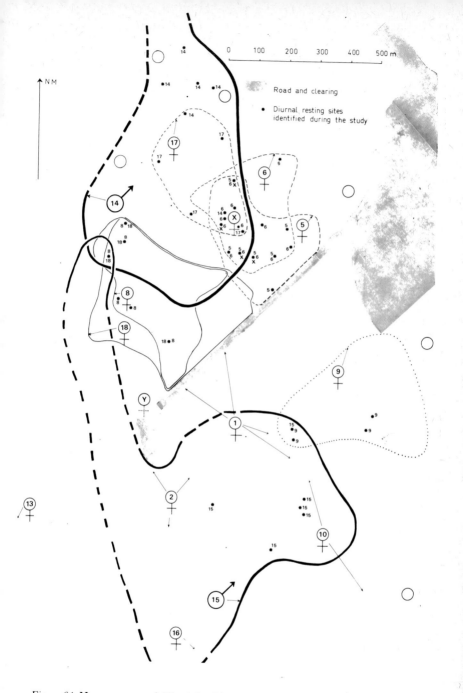

Figure 64. Home ranges of Allen's bushbabies followed by radio-tracking in primary forest in 1973.

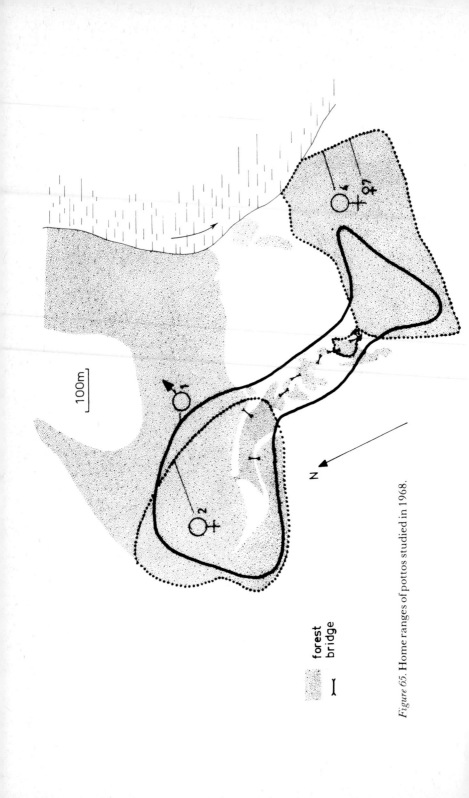

Figure 65. Home ranges of pottos studied in 1968.

forest
bridge

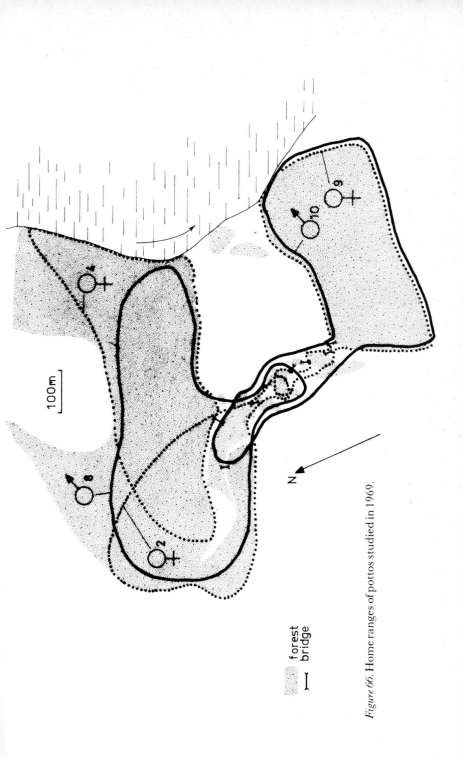

Figure 66. Home ranges of pottos studied in 1969.

forest
bridge

Table 8 Home range sizes for three of the lorisid species in Gabon

	♀ home ranges	♂ home ranges
Galago demidovii 60 g	0.8 hectares (range: 0.6 — 1.4) N = 6	0.5 — 2.7 hectares N = 6
Galago alleni 270 g	10 hectares (range: 8 — 16) N = 6	30 — 50 hectares N = 2
Perodicticus potto 1 kg	7.5 hectares (range: 6 — 9) N = 5	9 — 40 hectares N = 5

(i) *Female home ranges*

Within each species, the size of the home ranges of females is fairly constant, as compared to those of males, and it would seem that for females the size of the home range is broadly adjusted according to their basic needs. As noted above, the forest is far from uniform in constitution, and some regions seem to be 'better' than others. For example, with the population of pottos studied in secondary forest it was observed on two separate occasions that a female occupying an area adjacent to a 'good site' (a high clump of trees surrounding a spring) moved in to use this site when the resident female disappeared. In fact, the three different females which occupied this zone successively over seven years of observation always exhibited smaller home ranges (doubtless because of the higher local food availability) than when they occupied marginal areas lower down in the secondary forest.

Home ranges are continuously and thoroughly visited by their occupants. According to the season, some parts of the range are visited daily whilst others are left unvisited for longer periods. However, the interval between successive visits to any part of the range is never more than a week. (With *Galago alleni*, use of radio-tracking techniques has recently permitted verification of this observation with great accuracy.)

The reciprocal relationships between female home ranges are subject to great variation. Whereas the home ranges of adult female pottos only overlap to a very slight degree, all levels of overlap are found among the bushbabies. It will be seen later (p.241) that with the bushbaby species certain adult females exhibit association, often sleeping in groups with their offspring. With such females, the home ranges overlap extensively and sometimes completely. These females exhibit social contacts with one another, but never with females belonging to neighbouring groups. This therefore represents an incipient form of gregarious behaviour, which only clearly appears at the moment when the animals return to sleep. When the females awake again the next night, they resume their solitary activity.

It will be shown (p.244) that these groups have a maternal origin and that the tolerance towards one adult female exhibited by another is subject to individual variation in addition to the prevailing social conditions. Some relatively intolerant female bushbabies live alone, and this is not radically different from the situation characteristic of all lorisines. The particularly solitary pattern of life found with adult female lorisines must be related to their system of locomotion. We have seen above that the slow, silent movement exhibited by members of this subfamily represents a protective mechanism which avoids arousing the attention of predators. In consequence, all of the adaptations found in lorisines (e.g. dietary adaptation for feeding on slow-moving, repulsive and easily-detected prey) are developments towards greater economy of movement. By occupying in isolation a relatively small home range a female potto finds a greater concentration of available food than if she had to share a common, larger area with several other females. Her locomotion is accordingly restricted still further.

On three occasions in the period 1966-73, the actual adjustment of female home ranges was observed with the potto. On the first occasion, the two females concerned were a mother and her accompanying 10-month-old daughter. The mother moved 200 metres away, leaving her previous home range to her daughter. In the other two cases, the situation involved two adult females living in the vicinity of a zone with a clump of high trees surrounding a spring (see above). On each occasion, the female occupying this zone around the spring disappeared, and the neighbouring female moved into this zone (which was doubtless richer in fruits).

This occasional re-adjustment of female home ranges is not peculiar to the potto. The same phenomenon has been observed on some occasions with Demidoff's bushbaby. In the population studied over the period 1968-73, home range shifts of 100 and 200 metres (respectively) were observed in two instances with a total of 13 females followed from year to year.

(ii) *Male home ranges*

The sizes of male home ranges are determined more by the locations of females than by food-availability, and there is a general tendency for males to seek association with the maximum possible number of females. Since the sex-ratio in all five prosimian species is balanced at birth, it follows that there is active competition between males, doubtless giving rise to selection for the most vigorous among them. The smallest and weakest males are hence excluded to various degrees from contact with female home ranges.

A number of captured prosimians (5 *Perodicticus potto*; 3 *Galago demidovii*; 7 *Galago alleni*; 5 *Euoticus elegantulus*) were released in unfamiliar forest areas. The females generally exhibited rapid settlement in the new area, provided that it was not occupied by other females of the species and that it was ecologically appropriate. On the other hand, the released males moved on if no females were present, but remained in the area if females had already settled there. Further, if only females were released strange males soon appeared in the vicinity.

Table 8 shows that adult male *Galago demidovii* had home range sizes varying from 0.5 to 2.7 hectares, while male *Perodicticus potto* had home ranges from 9 to 40 hectares. With *Galago alleni*, the male home range may be as large as 50 hectares, compared to 8-16 hectares for females. For all three species, such variations in home range size are surely not attributable to the simple needs of foraging. For example, two male pottos (one adult; one immature, but weaned) were followed simultaneously by radio-tracking in the same primary forest area over the period 25.6.73 to 25.7.73. Whereas the young male obtained its food from a range 10 hectares in area (equivalent in size to that of an adult female), the adult male exploited a home range of 40 hectares in which every sector was visited at intervals of 1-4 days. It will be seen in the section on sexual behaviour (p.221) that the large male lorisines 'patrol' the home ranges of their females in a continuous fashion.

1. *Galago demidovii*. In this species, four categories of adult males can be distinguished on the basis of their body-weights and behaviour (cf. Charles-Dominique, 1972). 'Central A' males are the heaviest (average weight = 75 g) and occupy the central region of female populations, with which they exchange numerous social signals. The home ranges of these males are the largest observed (average = 1.8 hectares; maximum = 2.7 hectares). Two important characteristics can be noted with the disposition of these home ranges: first of all, if such a male is associated with only one female, the home range limits extend far beyond those of the female. However, if a Central A male is associated with several females, its home range is extended into a 'corridor' overlapping only partially with the female ranges concerned. Secondly, the home ranges of these large, central males converge at a common point (Fig.63), where there is a small degree of overlap. As will be seen, this overlap zone permits them to engage in reciprocal assessment of neighbouring males.

In 1969, in the population of *G. demidovii* under study ♂ 4 (whose home range during 1968 is represented in Fig.63) had disappeared and the entire range used by this male was absorbed by ♂ 9, which thereafter had an association with 5 females, rather than the single female available the year before. ♂ 8 had also disappeared by 1969, and the eastern sector of this male's former home range had been occupied by a new male weighing 75 g. ♂ 34 occupied the same range in 1969 as in the previous year.

The second category – 'Central B' males – is composed of the lightest males examined (average weight = 56 g), and these are tolerated in the central region by the large Central A males within this area of female home ranges. The Central B males have very small home ranges (approximately 0.5 hectares) and they have virtually no contact with the females.

Medium-sized 'Peripheral' males of the third category (average weight = 61 g) occupy relatively large home ranges adjacent to areas occupied by females. In their movements through these large home ranges (average size = 1.4 hectares), they may associate with other Peripheral males. However, their occupation of marginal ranges is only temporary. One male occupying such a range in 1968 ($\mathcal{J}$ 13 in Fig.56) was found to have shifted to the rank of a Central A male one year later. This male's body-weight had increased to 66 g from 60 g and the new home range occupied, 200 metres further into the population, was then found to overlap with the home ranges of two adult females.

Nomadic or 'Vagabond' males of the fourth category are in most cases young individuals which have already reached puberty. They do not stay very long in any one area. For example, during the period July-December 1968 the following numbers of Vagabond males were seen passing through the population under study: 2 in July, 1 in August, 1 in September, none in October, 2 in November and 2 in December. In contrast to the young females, young males leave their area of origin as soon as they reach puberty. Of 7 young males originally identified with their mothers in the study area between 1968 and 1972, not one was recaptured in the area after attaining an age of 10 months. This post-pubertal Vagabond stage doubtless ensures that exogamy takes place between neighbouring natural populations of *Galago demidovii*. The body-weights of young Vagabond males are extremely variable, and it is probably the largest among them which usually replace any Central A males which disappear.

The Central A males possess their central status only for a short period of time. Of eight such males originally recorded in the study area in 1968, only five were recaptured the following year, and only one was recaptured over the next three years. Their great activity, which exposes them to increased predation risks, and the fierce competition between them in the defence of associated female home ranges, doubtless combine to bring about rapid replacement.

2. *Galago alleni.* In Allen's bushbaby, the situation of the males is much easier to interpret than in Demidoff's bushbaby. In fact, as noted above (p.136), adult males are relatively few in number compared to females, and each male is uniquely associated with numerous female home ranges

within an enormous overlapping home range (Fig.64). The sex-ratio at birth (Table 5) would appear to be balanced, since hunting with a rifle yielded 6 young males and 5 young females. With the adults, on the other hand, only 4 males were obtained in this way, as compared to 17 females. Trapping in the study area gave a similar proportion of one adult male for every 5-6 females. Accordingly, there must be pronounced competition between adult males, such that most of them eventually disappear. This imbalance in the sex-ratio may be explained by the fact that the males, which move around a great deal, expose themselves more to predators and hence survive for shorter periods than the females (see p.221). In the study area in primary forest (Fig.64), the two adult males followed in 1972 had disappeared in 1973 and had been replaced by two new males, which had arrived from elsewhere. It should also be noted that the adult males are extremely aggressive towards one another. Following one fight observed in captivity, one male Allen's bushbaby died as a result of the injuries inflicted (open fracture of the femur).

3. *Perodicticus potto*. The social behaviour of this species has been described in detail in a recent publication (Charles-Dominique, 1974b). The males followed in the forest occupied home ranges of greatly varying size. The five ranges recorded were: 9 hectares, 12 hectares, 13 hectares, 15 hectares and 40 hectares. As with *Galago alleni*, there is an imbalance in the adult sex-ratio (15 males, as opposed to 27 females – cf. Table 5). Just as with the galagines studied, sedentary males whose ranges overlap with those of one or more females are somewhat heavier than other, 'Vagabond', males. With the noose-trap placed in the forest from October 1967 to April 1968, the following pottos were successively captured:

4.10.66 – juvenile female
6.10.66 – adult male
8.11.66 – adult female
31.12.66 – adult male
29.1.67 – adult male
3.2.67 – adult male
10.3.67 – adult male
24.3.67 – adult male
2.4.67 – adult male

The first three animals were probably sedentary occupants of the area, whereas the six adult males trapped subsequently were doubtless 'Vagabonds'. The average weight of the latter was 1010 g (range: 870 g-1155 g), whereas the average weight for sedentary males was 1120 g (range: 1100 g-1300 g, N = 12). However, all six of the apparent 'Vagabond' males had reached sexual maturity. The average weight of the testes for five such males was 5.4 g, as compared to 5.9 g for five other males shot during the same period (difference not significant).

B. SOCIAL COMMUNICATION

Before tackling the problem of social relationships exhibited by these prosimians, it is necessary to examine the means of communication possessed by the lorisids.

(i) *Visual signals*

The nocturnal, solitary way of life of the lorisids has prohibited them from developing complex postural signals comparable to those of the simians. However, in the course of contacts between individuals there are a certain number of postures – which may or may not be associated with vocalisations – permitting the reciprocal assessment of emotional states.

Submission is indicated among the bushbabies by folding of the ears, lowering of the head, passage beneath a branch, and (eventually) by straightforward flight. When a bushbaby is frightened but unable to flee, a crouching posture is adopted. The mouth is opened, the ears are spread widely, and in many cases the hands are held spread out in front of the body, whilst the animal utters a threat-call. This posture is neither submissive nor aggressive; it corresponds to the fright posture adopted by many mammals (arched back posture of the cat, etc.). With the lorisines, submission is accompanied by retraction of the head into the species-typical defence posture (between the arms for the potto; beneath the arm-pit for the angwantibo).

Among the galagines, aggression is indicated by an extended posture of the body and tail, with the ears widely spread and the mouth held open. At a somewhat higher level of arousal, the animal may utter a staccato aggressive call, and finally one hand may be lifted as a threat gesture. As a rule, this latter signal is sufficient to bring about the retreat of the antagonist, which will either adopt a submissive posture or flee. In some cases (*Galago demidovii; Euoticus elegantulus*), the dominant animal leaps abruptly alongside the antagonist, and the latter will either swing underneath the branch or leap downwards.[1]

The aggressive posture of the lorisines, with the exclusion of leaping, is entirely comparable to that of the galagines.

When a bushbaby moves along á support, the tail is held slightly raised at its distal end. However, when greatly excited by an unusual or alarming situation, *Galago demidovii* raises its tail in a question-mark form and then holds it parallel to the back, whilst uttering alarm-calls. This tail-position is not observed with the other two galagines. In addition, young *Galago demidovii* frequently adopt a peculiar posture with the tail coiled into a 'corkscrew' when interacting with adults. They are never attacked when in this posture, and when they are repulsed by an adult – following an over-intensive bout of play behaviour – the 'corkscrew' posture is rapidly adopted. The same tail-posture is found with *Galago alleni*, and with *Galago demidovii* (at least) it is doubtless an indicator of immaturity. (N.B. Kingdon (1971) believes that the orange/yellow coloration of the ears and the face of the young Demidoff's bushbaby is similarly an indicator of immaturity.)

Further significant postures are exhibited during allogrooming (Fig.67). An animal 'inviting' a partner to perform grooming lowers its head and stretches one arm forwards, thus presenting its neck and armpit region (Andrew, 1963). When allogrooming occurs between a male and a female, the latter may hang suspended beneath the branch by its hind-limbs. The male then adopts the same posture and the two animals, hanging head downwards and face-to-face, subsequently engage in mutual licking. This swinging beneath the branch during grooming is a prelude to

---

[1] With *Galago demidovii*, rapid drumming with the hands on the support has been observed, accompanied by slight withdrawal. It is difficult to interpret this 'display', which only appears occasionally in certain individuals, following excitation of some kind.

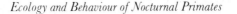

Figure 67. Allogrooming in the angwantibo. Note the positions of the arm and the neck, typical of all lorisids and numerous other primates when engaged in allogrooming. (Drawings prepared from an original photograph).

mating. The female *Galago demidovii* clings with all four limbs beneath the support when ready for mating to take place.

(ii) *Vocal signals*

Certain powerful vocal signals permit communication over considerable distances, whilst other, weaker calls are involved in short-distance interactions. In the latter case, they are usually associated with the postures which have been described above. Whereas the galagines exhibit a great variety of vocal signals, the lorisines only possess weak or moderately-developed calls which are utilised over a short range in the course of their rare encounters. Since the protective mechanisms of the lorisines reside essentially in discretion and concealment (see p.86), the absence of powerful vocal signals likely to arouse the attention of predators is readily understandable. Beaudenon (1949), in an account of *Perodicticus potto* from South Togo, and Manley (pers. comm.), with reference to *Perodicticus potto* and *Arctocebus calabarensis* probably originating from Nigeria, have described a powerful whistling call for the two African lorisines. However, with the two subspecies studied in Gabon by the author, such calls have never been noted either in the field or in captivity. It is hence possible that the lorisine species have generally lost their more powerful calls in evolution, but that certain populations (subspecies?) have retained some calls of this kind. As with most other mammals which have been studied, the calls are subject to a certain amount of variability which may indicate differences in emotional state; but among the lorisids generally the calls are relatively stereotyped.

An attempt has been made to establish similarities in call structure between the various species by study of sonagrams and by observation of the association between patterns of behaviour and certain vocalisations. However, study of the physical characteristics of these calls is still in its infancy, especially with respect to ontogenetic derivation, and any homologies between species which are suggested will probably require revision in the light of subsequent more detailed study.

A basic distinction can be drawn between *social contact calls* generally associated with intraspecific cohesion (classified as types A, B, C and D), *alarm calls* associated with excitation usually arising from disturbances in the animals' surroundings, *threat calls* concerned with avoidance of intraspecific contact or with the intimidation of nearby conspecifics, and *distress calls* uttered following injury during intraspecific fights.

1. *Galago demidovii*
(a) *Type A calls:* '*click*', '*tsic*' and '*gathering call*'. The neonate frequently utters 'clicks' consisting of extremely brief bursts of white noise (duration: 0.0005 sec) whose frequency covers a very large spectrum ranging up to ultrasonic levels beyond 40,000 Hz. The 'click' is also occasionally uttered by adults, particularly by females during mother-infant interactions in the nest and by males during courtship.

As the young bushbaby becomes more excited, the 'clicks' become closely aggregated in bouts and it can be seen that the energy is concentrated at various levels corresponding to harmonics of the incipient call thus formed (Fig.68A). At even higher levels of excitation (following a long period of isolation or in response to harsh handling) a typical short 'tsic' call (duration: 0.001 sec) is produced. The call is modulated in a parabolic form and the peaks of the numerous harmonics are located at the following levels: 8000; 15000; 21000; 28000; 35000; 41000; 55000; 64000 Hz (Fig.68). Since the recording of this call was conducted with a standard Nagra III tape-recorder and a Sennheiser MD 21 microphone, the ultrasonic components are evidently greatly attenuated and it is difficult to determine the level of maximum energy-concentration.

At the age of 2-3 weeks, the bursts of white noise tend to disappear and the 'tsic' (parabolic pattern), sometimes preceded by a small number of white noise emissions, is uttered in bouts of 2, 3 or 4 units with intervals of 0.05-0.10 seconds (Fig.68B). As the animal grows further, this call is gradually organised into regular series, which are eventually arranged in the adult as a sequence of 15-20 units organised as a crescendo lasting approximately 2 seconds (maximum: 4-5 sec) (Fig.68C, D). It is this vocalisation which is referred to as the 'gathering call'. In this crescendo call, the pitch of each unit is gradually increased as the intervals between units are progressively increased by small increments (Fig.68). The maximum energy output is concentrated in the last 4-5 units of the sequence, which varies from individual to individual such that the vocalising animal can be identified by this call if the observer has had time to become acquainted with the population under study. In a previous publication, this vocalisation was referred to as 'call III' or the 'gathering call' (Charles-Dominique, 1972), and it is doubtless the same as the first vocalisation listed by Vincent (1969: p.204).

In the forest, the 'gathering calls' carry over distances of up to 100 metres for the human ear and they are mainly heard just before dawn at the time when the individuals of a group are re-assembling in order to return to a single nest or to move to a sleeping-site in the foliage (Fig.68D). In the latter case, the site varies from day to day. During the re-grouping phase, these calls elicit further calls of the same kind, but vocalisation ceases as soon as the animals are reunited (in the space of a few minutes). On one occasion, when the author happened to

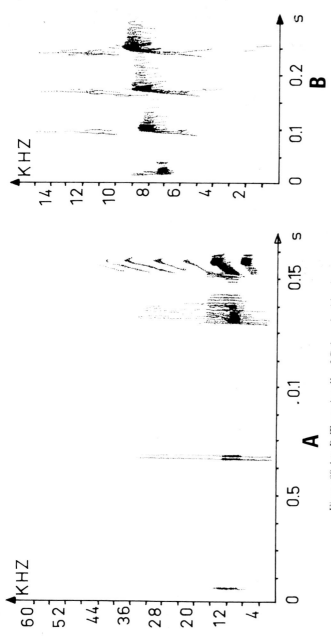

Figure 68 A to D. Type A calls of *Galago demidovii.*

A. 'Clicks' and 'tsic' produced by the neonate in one sequence. From left to right: isolated 'clicks' – groups of 'clicks' – initiation of a 'tsic' – complete 'tsic', representing a maximum degree of excitation of the neonate.

B. Call formed by a series of 4 'tsic' elements uttered in a crescendo by a young animal aged 2 weeks.

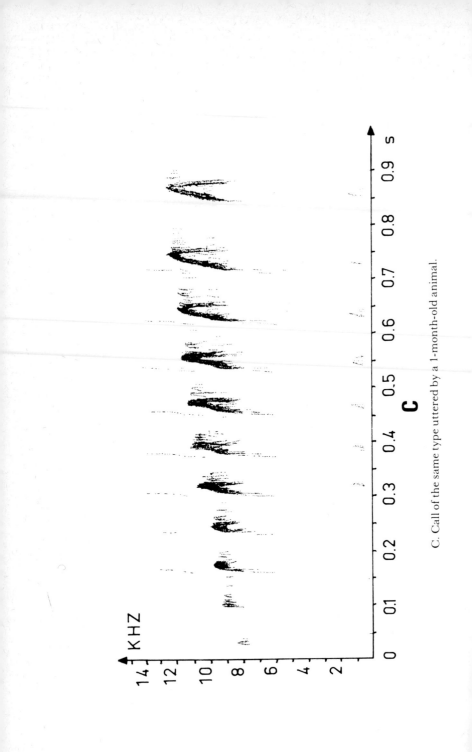

C. Call of the same type uttered by a 1-month-old animal.

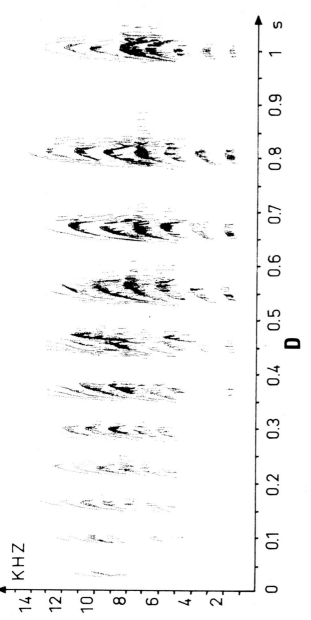

D. Call of the same type uttered by an adult (= morning re-grouping call). Only the latter part of the vocalisation is represented in the sonagram.

(Calls recorded with Nagra III tape-recorder at a speed of 38.1 cm/sec, with a Sennheiser MD 21 microphone. Under these conditions, ultrasounds above 20 KHz are extremely attenuated in recording.)

Note that in A the speed of analysis was divided by a factor of 4 in order to spread out the representation of the call. This has modified the scale of the abscissa.

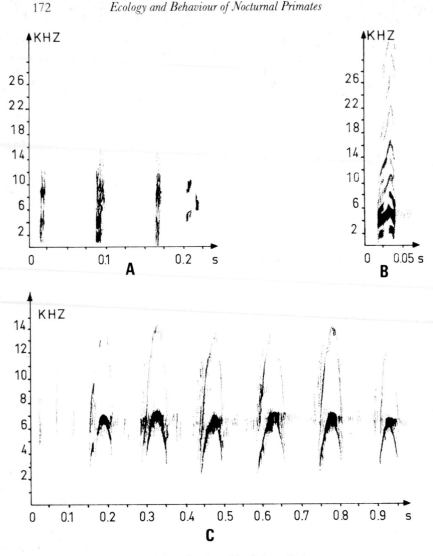

Figure 69. Development of the infant's call in *Galago alleni*.
A. 'Click' emitted by the neonate.
B. At a higher level of excitation, the same neonate utters a 'click' comparable to that of *G. demidovii*.
C. Agonistic call of the adult *G. alleni*, derived from the 'click' of the neonate.
    (Recording made with Nagra III tape-recorder at a speed of 38.1 cm/sec, using a Sennheiser MD 21 microphone.)
Note: In A and B the speed of analysis has been divided by a factor of 2.

be at the re-assembly point of 5 or 6 Demidoff's bushbabies in primary forest, 70 call sequences were counted within a period of 9 minutes. The 'gathering call' certainly has a double function: it ensures both attraction and localisation of individuals over considerable distances, thus facilitating the re-assembly of group members which have been dispersed during the night. This call may also be heard occasionally at the beginning of the night and during the night-time activity phase. However, under such conditions reciprocal calling is rare and no re-grouping results. In captivity, the 'gathering call' is uttered both at the conclusion of activity and at the time of awakening. The frequent calls at the beginning of the activity phase, contrasting with the situation in the wild, are doubtless a product of social disequilibriation brought about by conditions in captivity.

(b) *Type B calls: 'rolling calls'.* This call-type has a rolling, continuous character and only carries over a few metres. It is uttered by adults in a number of situations. When a mother is close to her infant, she produces this 'rolling call', which elicits a 'click' from the offspring permitting her to locate its position. The same vocalisation is used when bushbabies in a social group have been reunited just before dawn by the 'gathering call'. The reunited group moves towards the sleeping-site in single file, with the members uttering the 'rolling call' (no doubt in association with the need for cohesion). This 'rolling call' is also uttered continuously by any male seeking contact with a female followed at a distance of a few metres. In this latter case, the vocalisation may be progressively transformed into a 'gathering call' (Fig.70H). Usually, the female will stop after moving some distance and the two partners then engage in grooming. In all cases, the 'rolling call' ceases as soon as full contact has been achieved: when the mother has recovered her infant, when the individuals of a group are reunited in the nest, and when the male is engaged in grooming the female. This is, accordingly, a short-range call associated with adult contact-seeking behaviour.

(c) *Type C calls: 'plaintive squeak'.* Both males and females may utter a weak 'plaintive squeak' for periods of 1-10 minutes (80-120 calls/minute) (Fig.70I). The distance over which the call

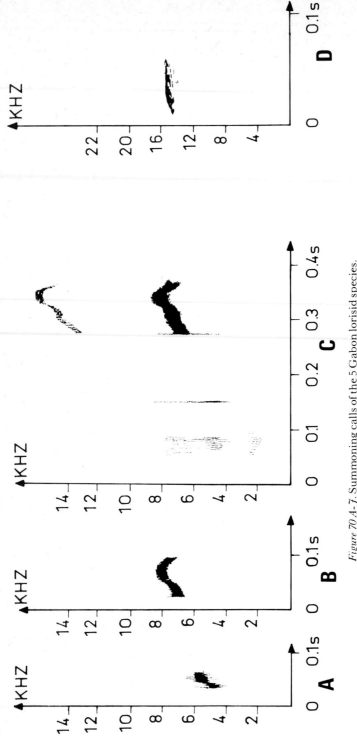

Figure 70 A-7. Summoning calls of the 5 Gabon lorisid species.
A,B,C. 'Tsic', calls of a young *Perodicticus potto*. At an intense level of
excitation (C), a slight groan appears before emission of the parabola,
which is longer and more acute.
D. Type D call of a ♀ *Perodicticus potto* in oestrus. This call may be
considerably prolonged.

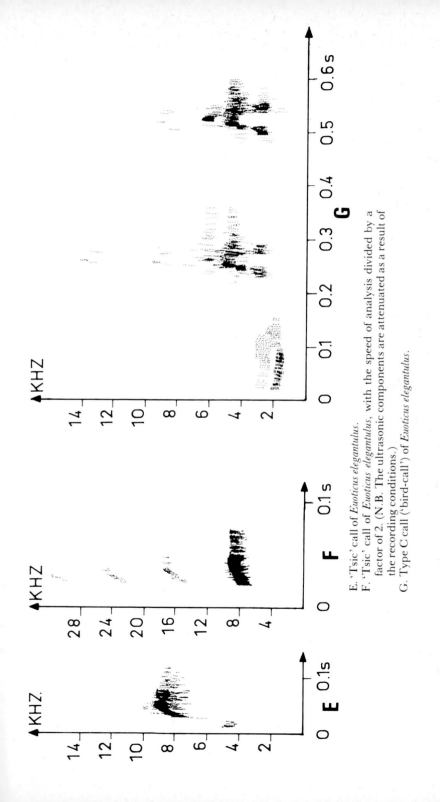

E. 'Tsic' call of *Euoticus elegantulus*.
F. 'Tsic' call of *Euoticus elegantulus*, with the speed of analysis divided by a
   factor of 2. (N.B. The ultrasonic components are attenuated as a result of
   the recording conditions.)
G. Type C call ('bird-call') of *Euoticus elegantulus*.

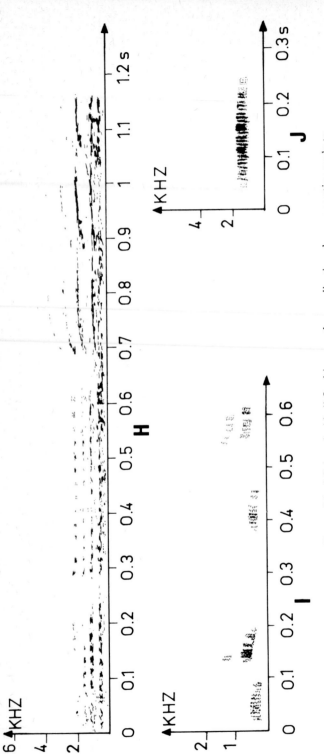

H. Type B call ('rolling call') of *Galago demidovii*. In this case, the vocalisation changes continuously to produce a rhythmic pattern, progressively giving way to the 're-grouping call', shown in its complete form in Fig. 68 (D).

I. Type C call of *Galago demidovii* ('plaintive squeak'). Two individuals are calling to one another at intervals of 0.1 and 0.2 seconds, with the second individual showing a somewhat more high-pitched squeak.

J. Type B call of *Galago alleni*. This vocalisation, whose energy is concentrated in a band between 1 and 2 KHz, is of a pure form and corresponds to the frequency band which is best transmitted in the forest undergrowth. This element is usually repeated a dozen times in each sequence.

(Recorded with Nagra III tape-recorder, at a speed of 38.1 cm/sec and using a Sennheiser MD 21 carries is variable, with a maximum of 60 metres or so (to the human ear) in primary forest. It has frequently been observed that these calls are associated with sexual excitation. When a pair separated for a period of several weeks is reunited, it is common to hear both the male and the female producing 'plaintive squeaks', and the call is also common around the time of oestrus (an isolated female may also emit this vocalisation during oestrus). When a female, as yet unreceptive, is being courted by a male, the latter frequently utters this plaintive call, especially in the nest where the female usually sleeps. These different observed contexts provide good grounds for believing that the 'plaintive squeak' is generally associated with sexual relationships, or at least with social relationships between males and females. Sometimes, *Galago demidovii* links this plaintive call with the 'alarm call' in such a way that the observer is given the impression that two different animals are vocalising. In such cases, without any obvious cause, several individuals may utter combined plaintive calls and alarm calls successively or in chorus.

(d) *Alarm calls: 'chips' and 'groaning call'.* When subjected to general arousal (by the presence of a predator or the human observer, by an unusual noise, by the introduction of a strange conspecific, etc.), *Galago demidovii* utters 'chip' calls whose frequency and structure change as arousal increases. These 'alarm calls' probably have multiple interpretations:

a – At an initial level of arousal, *Galago demidovii* utters a 'chip' followed by a weak 'groaning call'. The 'chip', which lasts about 0.035-0.050 seconds, exhibits frequencies modulated in a parabolic pattern, with an energy maximum situated at 5000-8000 Hz. (Fig.71A, B).

β – A little while later, the animal emits 'chips' in series of 3-4 units. The parabolic pattern is the same (fundamental peak at 8000 Hz; peak of the second harmonic at about 17000 Hz), but at the same time one can note the appearance of a structure intermediate between a chevron and a parabola, with the fundamental located at about 1000 Hz and several harmonics ranging up to

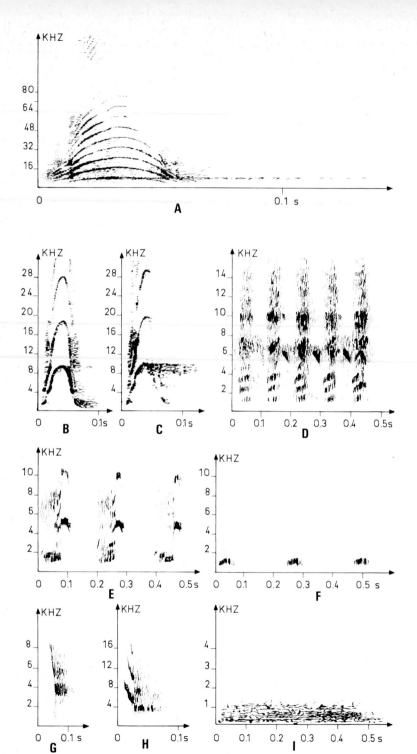

16000 Hz and above. From this point onwards, this call diverges in two different directions (γ and δ).

γ – This call consists of a rapid series of 8-15 'chips' each lasting about 0.05 seconds, with the individual units occurring at intervals of about 0.05 seconds (10-11 'chips' per second) (Fig. 71D). In this arrangement, the structure of the 'chip' is primarily based on the element intermediate between the chevron and the parabola (fundamental at 1000-1500 Hz) with several harmonics at 2500, 3500, 4500, 5500, 6500, 7500, 8500, 9500, 10500, 11500, 12500, 13500, 14500 Hz, etc. This call sequence may be repeated a number of times and in some cases it may lead on to the δ-type.

δ – The second direction in which the alarm calls may develop is in fact the most common, arising when *G. demidovii* is in a veritable paroxysm of arousal. In this case, the 'chips' are much shorter and they are composed of the ascending component of the parabola, with the energy concentrated between 4000 and 11000 Hz and the

Figure 71 (left). Alarm calls of the three galagine species.

A. Alarm call of *Galago demidovii*, at the a stage of initial excitation. The usual speed of analysis was divided by 8 in order to bring out the harmonics in the ultrasound region.

B. The same alarm call as in A, but with the usual speed of analysis divided by 2. Note the 'groaning call' (2-16 KHz), terminating the 'chip' in the form of a parabola.

C. Alarm call of *Galago demidovii* at a more advanced stage of excitation than that shown in A or B. It is the ascending segment of the parabola which is most accentuated.

D. Rapid alarm calls of *G. demidovii* at the γ stage.

E. Alarm calls of *Galago alleni* recorded at a distance of several meters from the source. Note the double structure of the call, with a chevron and parabola.

F. Alarm calls of *G. alleni* recorded in primary forest at a distance of about 60 meters. The high-pitched part of the call has disappeared, blocked by the undergrowth. Only the low-pitched part of the call (chevron) between 1 and 2 KHz is transmitted.

I. 'Groaning call' of *Galago demidovii*, preceding alarm calls in the initial stages of excitation. This 'groaning call' is only audible at a distance of a few meters, in contrast to the alarm calls, which carry over 100 meters.

(Calls A, B and C recorded with a Nagra IV SJ tape-recorder, adapted for recording of ultrasound. Recorded at a speed of 38.1 cm/sec with a B. and K. 4133 microphone. Calls D to I recorded with a Nagra III tape-recorder at a speed of 38.1 cm/sec using a Sennheiser MD 21 microphone.)

peak of the parabola located at 9000-11000 Hz, whilst a harmonic is present in the ultrasonic range (Fig.71). Rather than being organised in short sequences of 8-15 units, as with the γ-type, the 'chips' are uttered in a virtually continuous fashion at intervals of about 0.25-0.40 seconds, sometimes for as much as an hour. In primary forest, for the human ear, these calls carry about 100-150 metres.

Alarm calls are the most common category of the vocal repertoire of *Galago demidovii*, and they have been reported by a number of authors. Vincent (1969) records a 'third type of call elicited by height contrast stimuli (*sic*)', whilst Struhsaker (1970) has analysed 'chip calls' into a gradient of 'chevron', 'half-chevron', 'semi half-chevron', 'semi parabola' and 'parabola' types, with energy maxima situated between 8000 and 18000 Hz.

The initial stages of the alarm call $(a/\beta/\gamma)$, which reflect relatively mild arousal, are often uttered spontaneously during the first half-hour following awakening of *Galago demidovii* ('call I' of Charles-Dominique, 1972). At that time, the bushbabies of a group are in the process of dispersal and they exhibit reciprocal calls carrying 30-50 metres (to the human ear, and probably to the *Galago* ear as well). In the main phase of nocturnal activity, however, these calls are uttered less frequently. The function of such calls is doubtless that of signalling the respective positions of the animals concerned. The final stage of the alarm call (δ), on the other hand, is often associated with the presence of danger (mobbing, warning of the presence of predators and conditioning of the offspring to recognise predators). In other cases, alarm calls of type δ are found to follow territorial disputes, and they may occasionally be uttered spontaneously at certain points in the home range.

A series of experiments was conducted in captivity with the aid of a battery of cages with numerous compartments interconnected through doors with sliding partitions. An individual located in one compartment could hear the calls uttered by a conspecific in an adjacent compartment, but could only see the other animal by passing through the connecting door. If one animal (male or female) produced powerful calls of the δ-type, separated conspecifics would often come to join the vocalising animal if it were a member of their social group. If this was not the case, the other animals would remain in their own compartments. One should not, however, interpret the alarm calls as a precise signal eliciting a precise response. The calls are concerned with long-range communication, indicating the identity and degree of excitation of the vocalising animal. Conspecifics respond differently according to their social relationship with respect to that individual, their own location, their own level of arousal and a number of other elements determining their social status.

(e) *Threat calls: 'spitting vocalisations'.* When a *Galago demidovii* utters a threat call towards a conspecific, the latter will usually retreat or otherwise avoid contact. The threat call, which is audible over a distance of several metres, consists of a staccato series of 'spitting' sounds (Fig. 72F), produced with the mouth open and associated with a characteristic posture. When an animal of this species is restrained or cornered, its threat call is transformed into a hoarse growl with two components (inspiration and expiration). The same basic call is found with the other lorisids and with the cheirogaleines (mouse and dwarf lemurs) of Madagascar. Similar calls are found in almost all other mammal Orders.

(f) *Distress-calls: 'wheet calls'.* The distress-call, resembling that common in rodents, consists of a long whistling vocalisation of increasing frequency (Fig.72G). It is uttered by all animals following injury and sometimes simply during capture. In the course of fights between these bushbabies, if one of the protagonists utters a distress-call, the other will release its hold. Thus, this signal may limit the extent of injuries incurred in intraspecific fighting. As already mentioned, the local Gabonese village children knew how to imitate this distress-call by inhaling strongly through the mouth with the lips held firmly pressed together. With this device, they were able to attract the bushbabies in order to hunt them with a crossbow at dusk. The author, after learning this technique of imitation, was frequently able to attract various individuals in order to examine them at closer quarters

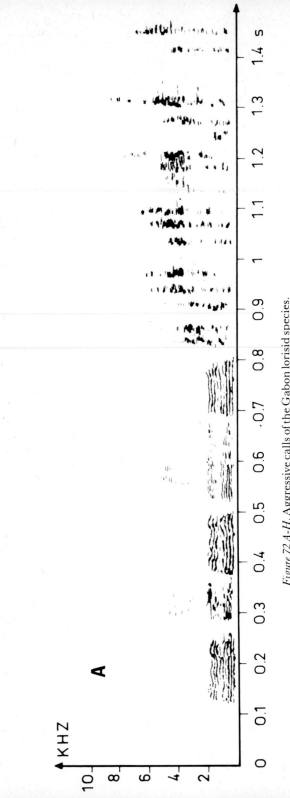

Figure 72 A–H. Aggressive calls of the Gabon lorisid species.
A. 'Two-phase groan' and 'threat call' associated in *Euoticus elegantulus*. The first part (two-phase groan) corresponds to a typical vocalisation uttered in a succession of expirations and inhalations. The second, more high-pitched, part is generally associated with aggression from a conspecific which has approached too closely.

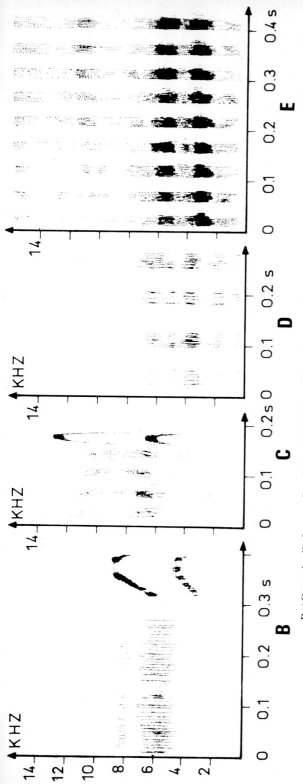

B. 'Groan' call of young *Perodicticus potto*, followed by a distress-call.

C. 'Groan' call of a *Perodicticus potto*, followed by a 'hee' call (associated with aggression from a conspecific or approach of a predator).

D. 'Groan' call of *Perodicticus potto* organised into series. The additional vocal contrast indicates increased aggression.

E. Paroxysm of aggression in *Perodicticus potto*; in addition to the vocal contrast produced by organisation of the 'groan' calls into series, a second accentuation is produced by increased energy in the high-pitched part of the call. Frequently, the 'hee' call (see C) follows this type of 'groan' call at the point where the potto makes lunges at the source of the menace evoking the call.

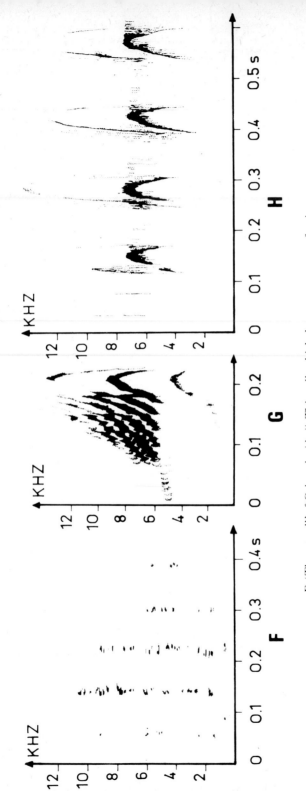

F. 'Threat call' of *Galago demidovii*. This call, which does not carry very far, is directed at conspecifics which approach too closely.

G. Distress-call of *Galago demidovii*. This call breaks off intraspecific fighting.

H. 'Threat call' of *Galago alleni*.

(All calls recorded with a Nagra III tape-recorded at a speed of 38.1 cm/sec, with a Sennheiser MD 21 microphone.)

(identification of tail-marks, etc.). All bushbabies are attracted in this way, but lactating females respond the most readily.

There are undoubtedly other vocalisations which are produced in the ultrasonic range, particularly in the nest where a barely audible gurgling sound can sometimes be heard. As far as the common calls, heard during the animals' activity periods, are concerned, four are of only weak intensity and utilised in the course of direct contacts: 'rolling call' (cohesion-seeking behaviour), threat call, weak 'click' (mother-infant relationships), and the distress-call. In addition, there are three long-range vocalisations: one has a special rôle ('gathering call'), one has a probable sexual rôle ('plaintive squeak'), and the third has a warning function (alarm call).

2. *Galago alleni*

(a) *Type A calls: 'click' and 'tsic'.* As with the young *Galago demidovii*, the young *Galago alleni* emits short 'clicks' when aroused (bursts of white noise lasting 0.0025 sec). As a rule, these calls are uttered spontaneously in the morning as the sun is beginning to rise. If the mother does not come to collect the vocalising infant, the 'clicks' change rapidly as the offspring becomes increasingly aroused, as with Demidoff's bushbaby. The burst of white noise leads on to a series of such bursts with the energy concentrated in the harmonics of the incipient call, and this in turn gives way to a 'tsic' call of parabolic form with numerous harmonics (fundamental peak at 3500 Hz; harmonic peaks at 6500, 9000, 12000, 14500, 17000, 22500, 28000, 32000 Hz, etc.) (Fig. 69A, B). The 'tsic' call, which lasts about 0.025 seconds, may be repeated in sequences of 2-6 units. The call is uttered when the infant is badly placed in the nest, with respect to its mother, and is seeking a more 'comfortable' situation. The mother's response is to allow the infant to move around until it has located a teat or has simply burrowed beneath her body. Once a suitable position has been adopted, these faint high-pitched calls cease to be emitted. As with *Galago demidovii*, the 'clicks' (bursts of white noise) uttered at an initial level of excitation suffice for location of the infant by the mother when she is searching in the vegetation.[1]

With increasing age, the parabolic 'tsic' call changes in a direction totally different from that found with *Galago demidovii*. The majority of the harmonics are lost and the resulting call is uttered when the animal is angered (cf. threat call (Fig.69C)). On the other hand, the unmodified short 'click' calls are still uttered in the adult stage when an animal is seeking to establish contact (e.g. females regrouping to sleep in a tree-hollow).

(b) *Type B calls: 'croaking call'.* This social contact call, a kind of low-pitched croak, would appear to be the homologue of the 'rolling call' (Type B contact call) of *Galago demidovii*. In this differentiation in form of comparable vocalisations one can identify clear-cut adaptation of the vocalisations of the two species as a function of the habitat typically occupied (canopy for *Galago demidovii*; undergrowth for *Galago alleni*).

In the primary forest of Gabon, Chappuis (1971) has demonstrated that in the undergrowth, for equal amounts of energy utilised, it is the lowest frequencies which travel the greatest distance. Investigation of bird-song revealed that all of the species in the undergrowth utilise frequencies lying between 1000 and 1500 Hz. Comparison with closely-related species occupying a more open habitat showed that the latter utilise a much wider band of frequencies. (The study involved 50 species inhabiting the undergrowth and 50 open habitat species.)

The case of these two bushbaby species is entirely comparable. *Galago demidovii* typically occupies the upper stratum of primary forest and has developed as a long-range social contact call a vocalisation of Type A. The energy is largely concentrated in the harmonics, which occupy a broad frequency band (1500-13000 Hz). On the other hand, *Galago alleni*, which typically inhabits the undergrowth zone, has

---

[1] *Observation conducted in primary forest on 15.1.68*: At 10.00 p.m., a young *Galago alleni*, a few weeks old, was found immobile in the foliage at 8 metres in a shrub. The distress-call of the species was imitated by the author and the mother arrived on the scene 10 seconds later. At 5-6 metres from the infant, she uttered a fairly low-pitched, rolling croak (Type B call). The infant at once responded with a series of 'clicks', whereupon the mother moved up alongside, grasped the flank of her offspring in her mouth and hurried away.

developed a long-range contact call of Type B (rolling call of low frequency) whose energy maximum is located precisely between 1000 and 1500 Hz (Fig.70J). This long-range contact call, which is audible over a distance of at least 250 metres, is usually uttered in sequences of 10-30 units, each 0.20-0.25 seconds in duration and separated at intervals of about 2 seconds. These call units are emitted with increasing intensity as the sequence progresses. However, when a mother is calling to an infant located nearby, only one or two faint call units are uttered.

When this call is recorded with a tape-recorder and then played back to the animals, it almost always evokes a vocal response. In the forest, following of individuals of this species with radio-tracking provided verification that the male and female can establish contact with one another in this way over a distance of 250 metres and thus succeed in joining up. (The male only meets up with his associated females at night, since he sleeps alone during the daytime.) The associated females (matriarchy) in a group also communicate with one another by using this call during the night, though they do not necessarily meet up as a result. In many cases, two 'rival' individuals (2 males or 2 females) will produce this call reciprocally across a home range boundary.

(c) *Alarm calls: 'Kiou-kiou-kiou' call.* This type of vocalisation which, like the Type B call already described, is adapted for long-range communication in the forest, is also specially constructed for sound-propagation in the undergrowth. The unit call ('kiou'), which is organised into sequences of 2-20 according to the individual and its degree of excitation, is entirely comparable with the unit alarm call of *Galago demidovii*, corresponding to an early stage in the development of the call (β), in which there is superimposition of a parabolic structure on chevron elements with numerous harmonics. However, the energy distribution is concentrated both in the fundamental of the individual chevrons (1000-1500 Hz) and in the fundamental of the parabola (4500-5500 Hz: Fig.71E). In the undergrowth zone of primary forest, the structure of the call is rapidly degraded as the distance from the vocalising animal increases. At about 60 metres, the component located at 4500-5500 Hz has been completely eliminated by the vegetation and only the components located at 1000-1500 Hz are retained (Fig.71F). As a result, an unfamiliar observer is led to consider the call heard nearby and the call heard at a distance as two distinct vocalisations. One might ask why the alarm calls of *Galago alleni* incorporate a clearly developed component which cannot carry very far in the undergrowth. Either the adaptation of the vocalisation is incomplete (e.g. representing an intermediate evolutionary stage), or the component located at 4500-5500 Hz plays a special rôle with respect to nearby conspecifics, such that the alarm call performs a double function.

The sequence of repeated alarm call units varies greatly from one individual to another and it is hence possible to recognise with ease certain individuals quite reliably on the basis of the number of units in a sequence and the intervals between the units. As a rule, a *Galago alleni* uttering alarm calls becomes progressively more and more excited. At first, the animal utters a low-pitched groan (Fig.71I) homologous with that of *Galago demidovii*, followed by alarm calls organised as one or two units. Half a minute later, the level of excitation is quite high and the sequences have assumed the duration and 'pattern' characteristic of the individual. Usually, alarm calls are uttered for a period of 5-15 minutes. The pattern of the sequences (Fig.71) perhaps provides conspecifics with information about the identity of the individual. It is possible that such a function might compensate for the degradation of the signal, of which only one component travels over long distances in the undergrowth, but this hypothesis has yet to be verified.

Alarm calls are elicited by any disturbance in the surroundings: sound produced by an unusual object, arrival of a predator, etc. However, Allen's bushbaby also utters this call spontaneously on waking and in the course of nocturnal activity, and it then simply reflects the arousal of the vocalising bushbaby. The rôle played by the alarm call is thus comparable to that found with the two other galagines.

(d) *Threat-call: 'quee quee quee' call.* This call, often preceded at low levels of excitation by a growling vocalisation with two components (inspiration, expiration), consists of parabolic elements uttered in sequences of 3-10 (sometimes even more), and it plays the same rôle as the threat-call of *Galago demidovii*. Once again, threat-calls are associated with an aggressive posture. Analysis of sonagrams (Fig.69C and 72H) indicates that this vocalisation is probably derived from the infant call uttered in a situation of discomfort (cf. Type A 'tsic' call).

(e) *Distress-call: 'wheet' call.* This plaintive call is comparable to the distress-calls of the other species examined, and it is uttered in response to pain. It acts to attract conspecifics.

Galago alleni thus makes use of only two long-range vocalisations: a 'social contact call' (Type B) and an 'alarm' call. The first call would seem to play a part primarily in sexual relationships (reunion of the male and female during the night) and reciprocal territorial signalling. In contrast to *Galago demidovii*, there is no powerful vocalisation homologous to the 'gathering call' emitted just before dawn. It should be noted that *Galago alleni* often sleeps singly in a tree-hollow, and even when there is a group of females and offspring it never exceeds a total of four individuals, which find their way back to well-known localities without requiring a powerful vocalisation. When two individuals of this species arrive together in front of the tree-hollow used as a sleeping-site, they utter a number of low-intensity 'clicks' (Type A calls). The other, low-intensity vocalisations are generally comparable to those of *Galago demidovii*.

3. *Euoticus elegantulus*

The youngest individuals of this species examined during the study were already two months old and no 'clicks' equivalent to those of the neonates of the previous two species were noted.

(a) *Type A call: 'tsic' call.* The vocalisation produced by a young animal at the age of two months is roughly the same as that produced by adults. It consists of a short, high-pitched 'tsic' corresponding to the ascending component and the peak of a parabola, and it is for this reason that it has been classified as a Type A call. The base of the parabola is located at 7000 Hz and the peak at 9800 Hz, with harmonic peaks at

18000, 26500 and 33000 Hz (Fig.70E, F). The duration of the call is 0.035 seconds. 'Tsic' calls are uttered at the end of the night without any organisation into definite sequences. Generally, a needle-clawed bushbaby will utter one or two 'tsic' calls, and conspecifics respond with the same call, perhaps leading to further reciprocal calling. As soon as dawn begins to break, the animals belonging to a social group return to a definite zone in the forest where they habitually regroup before returning to their sleeping-sites. The positions of the latter vary from one day to the next, but all sites are located within an area of less than 100 metres diameter corresponding to a forest region which is particularly rich in gum-producing plant species. The social contact call uttered at the end of the night carries a distance of 60 metres to the human ear and doubtless further for *Euoticus elegantulus*, since the animals never take more than 5-10 minutes to regroup.

The same call is utilised during the night by the infant and its mother in order to re-establish contact.

(b) *Type B call: weak 'croaking' call*. A faint 'croaking' call, doubtless homologous with the 'rolling call' of *Galago demidovii* and the 'croaking' call of *Gallago alleni* is sometimes emitted by several *Euoticus elegantulus* when they began to utter 'tsic' calls. The 'croaking' calls rapidly disappear, leaving only the 'tsic' calls.

(c) *Type C call: 'bird-call'*. This call begins with a low-pitched groan, followed by 2-6 'quee' calls reminiscent of bird-song. The groan consists of a slightly descending chevron with the energy maximum situated at 1000-2000 Hz and harmonics initially located at 3000, 4000, 5000 and 6000 Hz (duration: 0.15 sec). The subsequent 'quee' calls, uttered at intervals of 0.10-0.35 seconds, are composed of parabolas (duration: 0.08 sec) with their bases located at 2000 Hz and their peaks at 4300. The peaks of the harmonics are located at 7500, 11000 and 14000 Hz (Fig.70G).

The groan uttered at the beginning of this vocalisation (produced by adults of both sexes) is probably homologous to the 'plaintive squeak' of *Galago demidovii*, and it is for this reason that it has been included in the same category (Type C call). It was first heard with a young female when aged 7-8 months, in a form close to that of the adult call but at a much weaker intensity. No information is available on the sexual behaviour of *Euoticus elegantulus*, which utters this call only on rare occasions as compared with the other vocalisations (similar to the situation with the 'plaintive squeak' of *Galago demidovii*). However, once an individual has started to utter this 'bird-call', vocalisation may last for 30 minutes.

(d) *Alarm call: 'tee-ya' call.* The most commonly uttered vocalisations of *Euoticus elegantulus* are the alarm calls. They may be transcribed as 'tee-ya', corresponding to the descending sweep of a parabola (= transient: cf. Fig.71G, H). The duration of the 'tee-ya' may be prolonged beyond the normal 0.035 sec if the animal is at a low level of arousal (onset of alarm calling). The energy of the call is distributed over the range 8000-3000 Hz. Alarm-calls are uttered in a continuous fashion, at intervals of 1.5-2.0 seconds, and a calling sequence may last 10-15 minutes. They occur in the same situations as with the alarm calls of the other two bushbaby species, and in primary forest they carry 150-200 metres to the human ear.

(e) *Threat-call: 'hom-han-ki-ki-ki' call.* The threat-call, which has a relatively short range (10-30 metres) begins with a two-phase groan corresponding to inhalation and exhalation. Two-phase groans of this kind may be repeated, but at high levels of excitation they are followed by faint, high-pitched noises (2000-6500 Hz) whose physical structure is difficult to decipher (Fig.72A). Threat-calls are uttered in the same situations as with the two previous species.

(f) *Distress-call.* This is comparable to that of the other bushbaby species, and it is uttered under the same conditions.

4. *Perodicticus potto*

(a) *Type A call: 'tsic' call.* The high-pitched, brief 'tsic' call is similar to that of *Euoticus elegantulus*. Sonagraphic analysis reveals that it has a parabolic form with the descending segment more or less truncated. Both the frequency and duration of the call vary according to the age of the individual and its degree of excitation (peak of the parabola between

6000 and 16000 Hz; duration of call = 0.033-0.120 seconds – Fig.70A, B). When this vocalisation is uttered by a very excited young animal which has been isolated for some time, as with *Galago demidovii* and *Galago alleni*, there emerge a number of extremely brief 'clicks' (white noise) preceding the 'tsic' characteristic of the species in general (Fig.70C).

This social contact call of the potto is primarily utilised by the mother and the infant just before dawn when they join up to move to a sleeping-site. On one single occasion in the field and on 3 occasions in captivity, this vocalisation was uttered by a male engaged in courting a female. In the forest, the call does not carry farther than 20-30 metres to the human ear.

(b) *Type D call: whistling call.* G. Manley (pers. comm.) has noted a whistling call, produced by both *Perodicticus potto* and *Arctocebus calabarensis*. Apparently, the same call is found with *Loris tardigradus* (J-J. Petter and C.M. Hladik, pers. comm.). The author has only on one occasion heard in captivity a very high-pitched whistling call of weak intensity (duration 0.5-1.0 second, frequency 16 KHz, Fig.70D), produced by a female *Perodicticus potto edwardsi* in oestrus. These whistles were produced at a somewhat variable rhythm, often with intervals of 20-60 seconds between them. In fact, it is possible that this vocalisation should be classified as a Type A call.

(c) *Threat-call: 'groan' and 'heee'.* This somewhat variable vocalisation is uttered in the course of intraspecific fights or encounters with predators. It is associated with the particularly elaborate defence behaviour of this species, which enables it to hold its own against small arboreal carnivores (see p.87). The 'groan' calls, which are initiated by a two-phase structure (inspiration, expiration), reflect a low level of aggressivity in the potto. At a somewhat higher level, the elements become intensified, more highly pitched (fundamental shifting from 2000-2500 to 2500-3500 Hz) and grouped together, producing a staccato groaning sound (Fig.72B, D, E). At the highest level of arousal, the staccato 'groaning' elements are followed by a high-pitched call ('heee') (Fig.72C). This call generally accompanies biting, or simulated biting, and often has an intimidating effect on a predator. It consists of a parabolic fundamental with a single harmonic, with the energy primarily concentrated at the peak of the parabola, as with the threat-call of *Galago alleni* (compare Figs.72C and 72H).

(d) *Distress-call: 'wheet' call*. This call consists of a high-pitched whistle of parabolic conformation. The energy maximum is concentrated in the first harmonic. The distress-call, which reflects fear or suffering, is frequently accompanied by non-staccato 'groan' calls.

5. *Arctocebus calabarensis*

(a) *Type A call: 'tsic' call*. The 'tsic' call is high-pitched and, to the human ear, indistinguishable from the homologous social contact call of the potto. When isolated, a young angwantibo will utter this call spontaneously as soon as the light intensity begins to increase, corresponding to the situation where the mother normally returns to collect her infant prior to moving to the diurnal sleeping-site.

(b) *Threat-call*. In the course of intraspecific fighting behaviour, or simply when they are handled, angwantibos produce a deep 'groaning' call with a two-phase structure (inhalation, exhalation), which may become more high-pitched and intense as the degree of arousal increases. These threat-calls are associated with the species-typical defence posture.

(c) *Distress-call: 'wheet' call*. This is basically the same as the distress call found in the other four prosimian species in Gabon.

In order to facilitate comparison of the various vocalisations between the five prosimian species, the different calls identified so far are listed in Table 9.

(iii) *Olfactory signals*

It is difficult to estimate the importance of the part played by olfaction in intraspecific communication between the prosimians under natural conditions. However, one can at least recognise two basic means of olfactory communication:

1. Indirect signals. The marking substance is deposited

Table 9 Summary of the calls of the five prosimian species

Vocalization category	Galago demidovii	Galago alleni	Euoticus elegantulus	Perodicticus potto	Arctocebus calabarensis
Type A call	"click", "tsic", "gathering call"	"click", "tsic"	"tsic"	"click", "tsic"	"tsic"
Type B call	"rolling call"	"croaking call"	weak "croaking call"	—	—
Type C call	"plaintive squeak"	—	"bird-call"	—	—
Type D call	—	—	—	"whistling call"	—
Alarm calls	"groaning call", "chip"	"groaning call", "kiou-kiou-kiou"	"tee-ya call"	—	—
Fright call (2-toned groaning call)	2-toned groaning call	2-toned groaning call	2-toned groaning call ("hon han")	2-toned groaning call	2-toned groaning call
Threat call	"spitting"	"quee quee quee"	"ki ki ki"	"rrriiii..."	—
Distress-call	"woo-ee"	"woo-ee"	"woo-ee"	"woo-ee"	"woo-ee"

Note:
Calls which are not underlined do not carry far in the forest. Single underlining indicates short-range carriage, double underlining indicates medium-range carriage, and triple underlining indicates long-range vocalisations.

and the odour will only be perceived later on by conspecifics moving past the marking-site. This would include urine-marking and – in the case of the lorisines – marking with scrotal and vulval cutaneous glands.
2. Direct signals. These are directly perceived at close quarters, during actual contact, and essentially involve intrinsic body odours.

1. *Indirect olfactory signals*

(a) *Urine-marking.* Marking with urine must be regarded as a complex form of communication incorporating multiple signals, whose significance cannot as yet be fully analysed. It is well known for most macrosmatic mammals (i.e. mammals with well-developed olfactory organs) that urine indicates to conspecifics the identity of the individual involved in deposition, its sex, and ⟨for females⟩ the phase of the ovarian cycle. Catabolic (breakdown) products of steroid hormones are involved as olfactory signals in the urine, alongside other signals produced by the accessory glands of the urinary system. The chemical nature of the olfactory signals involved and their pathways of action are virtually unknown, and most work in this field has been conducted with rodents rather than with primates. In the general realm of olfactory communication, very little work has been conducted with lorisids, or with prosimians in general. Yet olfactory signals are probably just as important as vocal signals in the behaviour of the bushbabies, and they are probably preponderant in the lorisines, whose social communication is almost entirely indirect.

To describe olfactory marking behaviour is equivalent to describing the manner in which an animal vocalises without stating the nature of the vocalisation actually emitted. Yet this is virtually what we are forced to do, in the absence of analysis of the content of olfactory signals.

It has, nonetheless, proved possible to verify the fact that the lorisids are able to recognise from a urine mark the identity of the individual that deposited it. Urine marks evoke much more intensive investigation when deposited by an unknown conspecific than when a familiar animal is involved. This has been observed on numerous occasions by allowing *Galago demidovii* in captivity to enter a new cage compartment.

If the compartment is unmarked, any bushbaby given access will enter rapidly and deposit urine. If an individual belonging to the same social group has previously deposited urine there, the response is the same. However, if the empty compartment has been marked by a strange animal, there is a long period of hesitation before entry, and sometimes the compartment will not be entered at all. (Bushbabies of low social rank hesitate the most.)

Seitz (1969) has experimentally demonstrated that the urine of the female potto is most attractive to the male when she is in oestrus. Thus, urine must play the same rôle as in most macrosmatic mammals studied to date.

In the absence of effective techniques for the analysis of transmitted olfactory signals, all that can be provided at present is a list of the ways in which each species disperses these signals within its home range, and an attempt at interpretation.

Galago demidovii and *Galago alleni* exhibit the 'urine-washing' pattern already described for *Galago crassicaudatus* (Eibl-Eibesfeldt, 1953), *Galago senegalensis* (Boulenger, 1936; Lowther, 1940; Sauer and Sauer, 1963), *Loris tardigradus* (Hill, 1938; Ilse, 1955), *Microcebus murinus* (Petter, 1962 and Manley, in Andrew & Klopman, 1974) and certain South-American cebid monkeys such as *Saimiri sciureus* (Hill, 1938) (see discussion by Andrew and Klopman, 1974). The pattern is concerned with moistening of the plantar surfaces of the hands and feet with urine. This is achieved by a one-sided movement in which the animal holds a support with its hands and one foot (e.g. the left) and brings the other foot (i.e. the right, in this case) beneath the penis or clitoris. Urine is expelled drop by drop on to the plantar surface of the foot, whilst the fingers of the hand on the same side (i.e. the right) are rubbed in the urine. After one or two seconds, the animal changes sides to moisten the hand and foot of the other side of the body, using the same technique in mirror-image form. In captivity, *Galago demidovii* marks about four times an hour in this way, the frequency increasing when the animal is very excited.

Sometimes, *Galago demidovii* and *Galago alleni* urinate directly on supports. In this case, they remain immobile whilst brushing the penis or the clitoris along the support and depositing only a few drops of urine.

Euoticus elegantulus, unlike species of the genus *Galago*, never practises 'urine-washing'. Urine is generally deposited directly on the support in small amounts. To do this, the animal lowers the rear end of its body slightly (with the tail raised) and deposits urine when the penis or clitoris touches the branch. With males, the penis exhibits mild erection at this point.

In captivity, *Euoticus elegantulus* also urinates directly, in small fine jets, into the water-bowls, which may sometimes be soiled from a distance of 1 metre when the bowls are introduced into the cage. In order to avoid this, it is necessary to place the water-bowls high up in the cage, and even then some individuals still manage to urinate into the water. Sometimes, people passing close to the cage are sprayed in the same way, and this indicates that the animal is capable of precise orientation of the urine jet.

Perodicticus potto. Urine-marking by the potto has already been described by Rahm (1960) and Seitz (1969). Whilst moving along a branch, the animal lowers its hindquarters (with the tail raised), touches the support with the end of the penis or clitoris and deposits a trail of urine roughly one metre in length. The vibrissae extending beyond the tip of the penis and clitoris doubtless play a part in guiding this behaviour of the potto, which is performed preferentially on broad branches. Sometimes, the potto will deposit urine without moving forwards, raising and lowering the hindquarters in a jerky fashion.

Arctocebus calabarensis produces large quantities of urine, which falls to the ground in most cases. In captivity, it was observed that the angwantibos always urinated and defaecated at the same sites. In certain cases, a trail of urine is left on the branch, as with the potto (cf. Jewell and Oates, 1969).

'Urine-washing' by the bushbabies and 'trail-marking' by the potto have been interpreted by various authors (Eibl-Eibesfeldt, 1953; Sauer and Sauer, 1963; Seitz, 1969) as a means by which the animals concerned may find their way through the forest. In addition, Ilse (1955) and others have proposed that this behaviour has a territorial function. However, observations conducted under natural conditions in

Gabon have demonstrated that the prosimians are generally perfectly familiar with their respective home ranges and that only in exceptional cases do they return by the same pathway the same night Indeed, the variability of pathways used on successive nights is in most cases so great that it is doubtful whether the odour from a scent-mark would still be present by the time the passage concerned is used again.

As an experimental procedure, certain passage links in the forest (branches crossing a pathway) were cut down and replaced by lianes stretched from one side to the other, often at a distance of more than 10 metres from the original 'bridge'. Nevertheless, the prosimians involved (*Perodicticus potto* and *Galago demidovii*) immediately made use of the new links, which indicates that sight is involved at least as a major contributory sense in pathfinding through the forest (cf. the preceding discussion of locomotion in the lorisines – p.70).

It is, of course, possible that natural odours emanating from certain trees in the forest also facilitate orientation within the home range. However, it is the author's opinion that the various behaviour patterns involved in dispersion of urine-marks represent adaptations to arboreal life ensuring that *conspecifics* are highly likely to encounter the deposited urine. Prosimians live in a three-dimensional realm. If urine were to be deposited in large quantities at a small number of sites, most of the fluid would fall to the ground and the resulting marking sites would only be rarely encountered by conspecifics, which similarly select from a wide range of possibilities in their movements through the fine branches. A bushbaby, for instance, is far more likely to encounter a urine-mark deposited in the form of a long trail left by the passage of another bushbaby than to come across a circumscribed spot of urine.

Galago demidovii and *Galago alleni*, which both utilise a great variety of supports in their movements through the forest (fine branches and lianes in the case of the first species; bases of small trees and lianes for the second) exhibit the greatest dispersion of urine-marks. *Euoticus elegantulus*, by contrast, moves primarily over broad branches and large-diameter lianes, which represent less varied means of passage and which are accordingly often visited by various conspecifics. Thus, the probability of discovery of urine-marks by conspecifics is relatively high. It has already been noted that gum-producing trees are exploited on the basis of regular circuits (p.41). Hence, scent-marks do not need to be so dispersed as with the other species, which utilise more varied pathways in an irregular fashion. In addition, the urine does not fall to the ground, even when deposited in a straightforward fashion, because of the large diameters of the supports used. (Unfortunately, it has so far proved impossible to determine in the field whether needle-clawed bushbabies urinate into small pools of water retained by branch-forks, as is suggested by the behaviour observed in captivity involving direction of jets of urine into water-bowls.)

The system of marking exhibited by the potto would seem to be clearly linked to its typical pattern of progression. Because of its means of locomotion involving exclusive climbing, it is necessary for this species to utilise successively in its passage from one tree to another fine branches and then much less numerous large-diameter branches (cf. discussion of locomotion on p.70). The potto only deposits urine-marks on the large-diameter branches, where the probability of passage of a conspecific is much higher.

Urine-marking plays an important social function by indicating to conspecifics the identity of any given animal, which deposits such marks throughout its home range (particularly in boundary areas). This system of communication, as indicated above, is commonly found among macrosmatic mammals. All that the lorisids have exhibited is special adaptation of this common mammalian pattern by utilising modes of urine-deposition which increase the probability that conspecifics will encounter the marks. It should also be noted that urine may be deposited on a social partner. With bushbabies, when the animals are engaged in mutual grooming, the partner may be held with the hands covered with urine in the course of 'urine-washing'. Similarly, Epps (1974) has observed that the potto may moisten its hands with urine which is then deposited on the partner.

(b) *Marking with scrotal and vulval glands.* In all strepsirhine prosimians (i.e. Lemuriformes + Lorisiformes – cf. Table 1), the male exhibits a gland on the scrotum which is developed to varying degrees according to the species and which has been subjected to histological study in a number of cases (Montagna, 1962; Montagna and Ellis, 1959, 1960; Montagna and Lobitz, 1963; Montagna et al., 1961; Machida and Giacometti, 1967). In *Perodicticus potto*, the scrotal gland is very well developed, whereas in *Arctocebus calabarensis* and the three galagine species of Gabon it is only moderately developed. In all species, it consists of a series of ridges in the troughs of which there are openings of both eccrine and apocrine glands.

Females of these species exhibit homologous glands on the 'labia' of the vulva, the glands being most conspicuously developed in the potto.

In the males, when they are not aroused, the surface of the scrotum is contracted so that the ridges are pressed against one another, thus imprisoning the glandular product. When a male is excited, however, the surface of the scrotum is stretched to reveal the ridges, with the result that the accumulated glandular product may be forcibly expelled (Fig.73).

With *Galago demidovii*, certain males were observed to turn round to present their genital organs to the female during the prelude to mating. Nevertheless, it would seem that in this species at least (which has been the best studied from this point of view) the product of the scrotal and vulval glands is primarily deposited on supports (cf. Vincent, 1969). This is thus a characteristic form of marking behaviour, though the author's observations indicate that the odoriferous substance is particularly deposited during encounters between males and females. An excited animal will rub its anogenital region on the branch, and the partner will then move up to sniff at the deposited trace. When a female has marked a support in this way, the male will sometimes rub his cheeks in the substance as if to pick up some of the odour. This same behaviour pattern is also exhibited on occasions when the female has performed urine-marking.

Some individuals rub the anogenital region frequently on branches, rapidly soiling them with a black substance. It would seem that faecal matter may also be applied to supports in the course of such marking behaviour, as is definitely the rule with dwarf lemurs (*Cheirogaleus* spp.) of Madagascar (Petter, 1962).

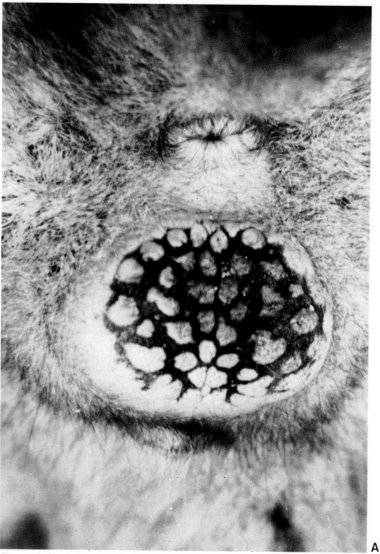

A

Figure 73. Genitalia of male (A) and female potto (B, overleaf), compared with the genitalia of male Allen's bushbaby (C, overleaf). (Photo: A.R. Devez)

For the lorisines, Manley (1974) has described for *Perodicticus potto* the manner in which the male, in the course of allogrooming with a female, scratches its scrotal gland with one hand and subsequently grasps his partner such as to

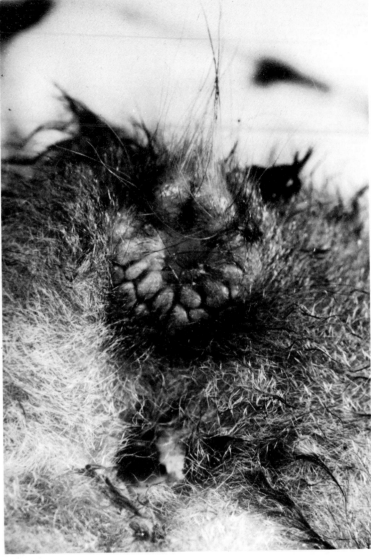

B

transfer the odour to her fur. The female potto behaves in the
same way, by scratching the homologous glands situated on
the vulval 'labia'. (This pattern was referred to by Manley as
'genital-scratching marking'.)

Manley (1974) has also described how the male *Arctocebus*

calabarensis rubs his scrotal gland on the female partner by straddling her ('passing-over behaviour') during allogrooming. In this case, the marking substances are deposited directly on the partner, as has been observed with lemurs of the family Lemuridae (Petter, 1962; Jolly, 1966).

Preclitoral glands of the potto. The female potto possesses a pair
of 'glands' of peculiar structure (Montagna and Yun, 1962;
Charles-Dominique, 1966b) situated on either side of the
clitoris. The 'glands' consist of pockets lined with folded
epithelium whose desquamation (sloughing) products
accumulate in the lumen and undergo considerable bacterial
degradation. There is no sudoriporous or sebaceous material
contained in the two 'glands', which are therefore not true
cutaneous glands. Each pocket measures about 5 mm by 3
mm, exhibiting no seasonal modification and becoming
functional even in young females. When a female is excited
(i.e. frightened), the musculature surrounding the pockets
contracts and expels through a small orifice a white, creamy
substance with a very repulsive odour. When a male isolated
in captivity is presented with a sample of this product, he
responds vigorously by running in all directions in the cage,
exploring all of the available space. In the course of 'courtship'
behaviour of the potto, observed on a single occasion in the
forest in Gabon, a female potto was seen to rub her genital
region on a branch. It is possible that the product of the
'preclitoral glands' is deposited at this time.

2. *Direct olfactory signals*

The general body odour of a prosimian plays an important
part in the course of inter-individual contacts. When two
animals meet, two body zones are subjected to particularly
intensive olfactory investigation: first of all, the cheeks and the
muzzle; secondly (especially for females), the genital region.
This olfactory investigation seems to be important, since when
two *Galago demidovii* are placed together in a cage with which
they are already familiar, their first action is to engage in
reciprocal muzzle-sniffing prior to initiating behaviour
appropriate to their social relationship (e.g. allogrooming,
separation, chasing, fighting, etc.). Of course, this only
applies when the two animals involved are accustomed to
living with conspecifics in captivity. If an encounter is
arranged between two bushbabies of which one does not
usually have a social companion, the latter animal will
immediately adopt either a submissive posture or an
aggressive posture (according to circumstances), which elicits
a response from the partner without any prior reciprocal
olfactory investigation.

It seems that individual recognition at very short range is hence necessary, at least for *Galago demidovii*.[1] The potto is the only lorisid to emanate a very strong (curry-like) odour. It will be seen in the section on relationships between neighbouring animals (p.212) that a conspecific can be discovered and recognised at a distance of 20-30 metres or more. In order to achieve this, pottos doubtless rely on their body odour, and it is possible that this also functions as a warning-signal to predators in association with the potto's vigorous defensive behaviour (see p.87).

The fact that the members of a given social group mix their respective odours by engaging in reciprocal marking doubtless provides complex indicators of the composition of the group to which each individual belongs. This has been verified on three separate occasions in captivity by isolating a male from his female towards the end of gestation. When returned to the female and her offspring one month after birth had taken place, the male became very excited and pursued both the mother and her young with equal intensity, licking them and attempting to mount (even though the offspring was male in two of the three cases). Only when the adult male's excitation had waned was his sexual behaviour oriented towards the mother. As with many other mammals, the female's genital region emits certain odours in all of the lorisid species. In the case of the bushbabies, the male often comes to investigate the female's genital region at the beginning of his night-time activity (see the discussion of male/female relationships on p.222).

Direct olfactory signals necessary for individual recognition can only play a part at short range (with the exception of the potto), operating in the course of inter-individual contacts which are common in the galagines and relatively rare in the lorisines. It would seem that glandular marking activity (particularly involving scrotal and vulval glands) has a comparable mode of action, since it occurs above all in the course of encounters between males and females and since

[1] In captivity, needle-clawed bushbabies recognise one another individually at a distance of several metres. Perhaps such recognition is based on individual differences in body-length and in the contour of the tail-tip, which is white. In a captive group, any intruder is immediately chased without any prior close-range investigation of the kind described above for captive *Galago demidovii*.

none of the species concerned utilises any means of dispersal for these glandular secretions. The latter would transmit information about the sexual condition of the marking animal to the conspecific partner present during glandular marking activity.

Marking with urine is, however, much more frequent than glandular marking. The mode of dispersal of urine-marks and the fact that they are not associated with a specific behavioural context indicates that the signals involved are destined for *any* conspecific which happens to pass through the home range. The signal has multiple information content, as indicated above, and conspecifics receiving the signal doubtless interpret it according to their social relationship to the marking animal. For a member of the same social group, this would involve indirect social communication; for a stranger, it would give an indication that the home range of another animal had been entered. Under natural conditions, an attempt was made to determine the behaviour of individual lorisids released in the home range of an unfamiliar conspecific (Charles-Dominique, 1974a). With *Euoticus elegantulus*, *Galago demidovii* and *Perodicticus potto* released in this way in the forest, it was found that both males and females immediately recognised that they were in the home range of a conspecific, and (without necessarily engaging in direct contact with the range occupant) the released animals moved on to establish themselves in an unoccupied zone. (To be more exact, females avoided the home ranges of other females and males avoided those of other males – see p.214.)

(iv) *Tactile signals*

Grooming of other individuals (allogrooming) plays an important social rôle in the lorisids. Cleaning is carried out by licking, and the 'tooth-scraper' (formed by the procumbent 2 canines and 4 incisors of the lower jaw as shown in Fig.19) is only involved when the hairs are stuck together or tangled. Loose hairs, which become wedged in the tooth-scraper, are removed by the sublingua (a flap of skin with cornified tips situated beneath the tongue) and subsequently swallowed. The lorisids – like all other prosimians, but in contrast to the simians – do not use their hands for actual grooming but merely to grasp the partner, whilst the grooming activities are carried out with the tongue and the tooth-scraper.

Social grooming is primarily exhibited by the mother towards her offspring, which is frequently licked by her. When the infant is 2-3 weeks old, it responds to the mother's grooming with reciprocal grooming activity. Thereafter, once the adult stage has been reached, allogrooming remains a frequent activity among individuals belonging to the same social group. In adult social grooming, the two partners adopt a face-to-face position and then take turns in grooming each other's face, scalp, ears, neck, armpits and shoulder regions. All of these are body areas which are inaccessible to any individual animal during self-grooming. Allogrooming of this kind may be initiated in the course of olfactory investigation of the facial region following an encounter between two individuals. In some cases, one partner may elicit allogrooming from the other by 'presenting' its neck and armpit with the arm stretched out in front (Fig.67). Allogrooming doubtless plays an important rôle in the reinforcement of social bonds, involving both tactile and olfactory stimulation between the partners. Actual care of the pelage, which is of great importance in young animals, is evidently of relatively minor importance with adults, since when they are isolated in captivity those regions of the body inaccessible to self-grooming remain identical in appearance to those of animals which are able to engage in allogrooming. Hence, the social rôle of allogrooming is predominant over the simple toilet function.

In bushbabies, allogrooming is particularly frequent at times when the animals are waking up or returning to their sleeping-sites. When a male joins up with a female in the course of the night-time activity period, they sometimes engage in allogrooming for several minutes. With the lorisines, which only exhibit rare encounters between individuals, allogrooming is practised between the mother and the infant and between the male and the female in association with courtship.

With bushbabies there is another category of tactile stimulation with a social rôle involved in groupings at nest-sites, where several individuals spend the day rolled up in 'contact groups' in the foliage. (This occurs in some cases with

Galago demidovii and in all cases with *Euoticus elegantulus*.) In the forest, the author has observed nesting groups of needle-clawed bushbabies containing up to seven individuals intimately entwined together and forming a bundle of the size of a football with their heads emerging at various points.

C. SOCIAL RELATIONSHIPS

Having examined for each of the five lorisid species the manner in which their home ranges are arranged with respect to one another and the ways in which the animals can communicate among themselves, we can now tackle the problem of social relationships. It should be remembered in this connection that the lorisids are social but solitary. That is to say that, despite the intercommunication exhibited among members of any given species, movements of individuals through their home ranges are not coordinated into group ranging activity of the kind found with the monkeys or with the diurnal Madagascar lemurs (Jolly, 1966). These latter species are gregarious and exhibit coordination of ranging over areas which may be of considerable size.

(i) *Relationship through adjacent ranges*

'Relationship through adjacent ranges' is taken to mean any relationship which is not based on membership of a given social group but on contiguity or slight overlap of home ranges occupied by individuals which do not form sleeping groups together.

1. *Perodicticus potto*

The slow-moving potto proved to be the best subject for observing the behaviour of neighbouring males in the zone of overlap between their home ranges. The potto population living in secondary forest, in the immediate vicinity of the former laboratory of Makokou, has been followed since 1966. This forest zone is transected by several roads, which divide it up into separate forest sectors inhabited by the pottos. In order to move from one such sector to another, the pottos made use of certain branches which served as bridges over the roads at certain points well-known to the animals. These

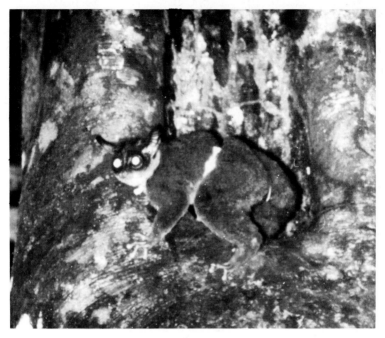

Figure 74. Photograph of an Allen's bushbaby carrying a radio-transmitter weighing 25 g, with a transmitting life of 5 weeks. The animal is just emerging from a tree-hollow in which it had been sleeping.

branches were progressively cut away and replaced by lianes stretched from one side of the road to the other. The pottos subsequently made use of these new 'bridges' and the lianes were accordingly fitted out with electrical contacts which permitted recording of the time and direction of any movement by a potto across them (see Charles-Dominique, 1974b). It was possible to follow changes in the various home ranges from one year to another, and in 1969 the situation happened to be particularly favourable for study of relationships between pottos with adjacent home ranges (Fig.66). The common area studied was the overlap zone between the home ranges of a pair of pottos located in the north (♂ 8 and ♀ 2) and the home ranges of a pair located further south (♂ 10 and ♀ 9). A single pathway, represented by a liane, permitted each of the two males (and each of the two females) to gain access *through their respective ranges* to a sector 0.5 hectares in area which constituted the major part of

Figure 75. Diagram of the device for recording activity of pottos in the study area near the earlier laboratory at Makokou. The animals moving from one 'island' of trees to another were obliged to cross liane bridges bearing electrical contacts. (Drawing prepared by Catherine Muñoz-Cuevas)

the common ranging area. (All other pathways into this sector of the forest had been eliminated by cutting down the branches involved.) On the two lianes which were left to provide access to the common area, electrical contacts linked to a bell in the laboratory immediately indicated any arrival in, or departure from, that sector of the forest. All four animals involved were marked and identifiable at a distance of up to fifteen metres. Thus, whenever the bell rang, it was fairly easy to move rapidly to the area and to identify the slow-moving pottos. In order to increase the likelihood of encountering pottos in this common area, a feeding platform baited with bananas was installed, and this was similarly linked electrically to the bell in the laboratory. Finally, in order to follow the pottos' activity in an uninterrupted fashion, the author reversed his own 'activity rhythm' in order to be ready to carry out observation at any time throughout the night.

Initially, it was expected that observations could be conducted on direct encounters in the vicinity of the artificial food-source; but the visits made there by the various pottos never coincided. In the course of 19 successive nights of continuous observation in the common area, the following visits were noted:

6 nights with visits from the north
7 nights with visits from the south
4 nights with visits from both the north and the south
2 nights with no visits to the common area

The females entered the common area essentially in search of food and they usually remained close to the artificial food-source. The males, on the other hand, scarcely ever visited the food-source and instead spent some time on each visit engaged in incursions into the neighbouring home range. Durations of 13 such incursions were noted as: 30 min. (3 cases); 50 min.; 1 hr; 2 hrs 30 min.; 1 hr 45 min.; 2 hrs 30 min.; 3 hrs (2 cases); 3 hrs 30 min.; and 4 hrs.

In fact, the pottos located conspecifics some distance away and exhibited mutual avoidance in the common area. On five separate occasions the author observed a potto arriving in the neighbourhood of the common area when the latter was already occupied by a potto from the adjacent range. On every occasion, the approaching potto stopped advancing towards the common zone, remained motionless for a period of one to two hours and then returned into its own range. For example, a male from one range will not venture into the common area if it is already occupied by the male or the female from the adjacent range. On the other hand, if the female from that male's range is already in the common area, he may move in to join her there. Similarly, the females will avoid encountering the neighbouring males, though they will move in to join the males from their own ranges. Only on one occasion were the two females seen to visit the common area on the same night (the night of 16 November 1969):

22.30 hrs: ♀ 9 seen eating at the artificial food-source
22.55 hrs: ♀ 9 retreats 50 metres to the south
01.10 hrs: ♀ 2 arrives in the common area from the north
01.35 hrs: ♀ 2 seen eating at the food-source
01.40 hrs: ♀ 2 moves out of the common area
02.38 hrs: ♀ 9 returns to the food-source

Thus, it would seem that the pottos can recognise one another individually from a distance, which may extend up to 50 metres at least, as indicated by observations in the field in Gabon. However, the manner in which such recognition is achieved remains a mystery. No relevant vocalisations were ever heard under field conditions, though one of the potto's vocalisations (type D call) is so high-pitched that in captivity it is only audible at a distance of several metres (Fig.70). It is possible that this call may carry far enough – to the potto's ears – to serve the function of long-range localisation. It is also possible that the strong odour emanating from the potto's fur permits recognition from some distance away. Yet it should be noted that a dozen pottos (both males and females) installed in wire cages only five metres away from the artificial food-source exerted no apparent effect on the behaviour of the freely circulating pottos. All that happened was that, in the first few nights following installation of the animals in cages, the resident pottos came to visit their captive conspecifics and to mark their cages with urine. Thereafter, they exhibited no further interest.

During their 'incursions' into the common area, it is possible that the pottos marked their presence with urine deposits. It was impossible to investigate this question in the field, since the pottos could only be observed from a distance. But in captivity both males and females distribute large quantities of urine on the separating cage-wire partitions between their cages (Seitz, 1969; author's personal observations). By means of such reciprocal marking, each individual can recognise its immediate neighbours, confirming their presence and informing them of its own presence.

2. *Galago demidovii*

A similar situation has been observed with *Galago demidovii*, in that all of the 'central A' males in the study zone (i.e. males with home ranges overlapping those of females involved in social relationships with them) made use of a common area. At three different capture points located in this area, common to four 'central A' males (cf. Fig. 63), the following numbers of captures were noted: ♂ 4 = 1; ♂ 8 = 3; ♂ 34 = 3; ♂ 9 = 1. The males were never captured together in any combination and only one male was captured (at the most) on any given night. The following minimal intervals between captures of different males were noted: 1 day between ♂ 4 and ♂ 8; 1 day between ♂ 8 and ♂ 34; 2 days between ♂ 4 and ♂ 9; 1 day between ♂ 9 and ♂ 8. This common area of approximately 1500 square metres did not seem to exhibit any ecological peculiarity with respect to surrounding forest zones. Thus it would seem that the neighbouring 'central A' males inform one another of their presence by means of urine-marking in this common area. This interpretation is supported by observations conducted in captivity with male *Galago demidovii* made to pass, one after another, into a common cage compartment.

3. *Galago alleni*

Allen's bushbaby exhibits the same behaviour as Demidoff's bushbaby. In primary forest, in the study area selected (Fig. 64), two males were found to share a small common area. Using radio-tracking (Fig. 74), ♂ 15 was followed for 15 nights and was only found to enter the common area in the course of one night, on which occasion the neighbouring male (♂ 14) remained at a distance of 50 metres, on the edge of the common area. In fact, ♂ 14 was only found in the common area three times in the entire period from 24 June to 17 July 1973. Observations were not carried out in a continuous fashion with each individual, but visits to the common area must surely be rarer than with Demidoff's bushbaby or with the potto. This is perhaps associated with the fact that male Allen's bushbabies cover enormous home ranges overlapping with those of 6-8 females (or even more). The powerful calls uttered by this species, both males and females, carry over a considerable distance in the forest. They very probably permit the various individuals to inform one another of their presence.

Neighbouring females may also exhibit common areas of ranging. For example, Fig.64 indicates a very slight overlap between the home ranges of ♀ 5 and ♀ 18. Yet these two females belonged to two different social groupings: ♀ 5 + ♀ 6 and ♀ 8 + ♀ 18, both related to one male. In the course of 45 days of observations, there were 2 separate occasions on which a calling concert ('croaking') was produced for a period of more than one hour by all 4 females, which responded to one another from their respective sides of the boundary zone with a distance of 20 metres separating the two neighbouring groups.

Relationships of this kind between occupants of neighbouring ranges are generally pacific and purely concerned with exchange of information. It is only when one is comparing one year with another that appreciable changes in location of home ranges emerge as a resultant of rare, periodic adjustments which are followed by long intervals of reciprocal surveillance.

(ii) *Relationships with strange conspecifics*
(Vagabond males)

It will be seen that females are sedentary from a very young age, whereas young males pass through a vagabond stage at the time of puberty. This mechanism, which ensures that exogamy occurs, is common to numerous mammal species; it permits young males to establish themselves over 'vacant' female home ranges or perhaps to displace the resident male. The fact that it is always the largest males which 'control' female home ranges gives rise to the suspicion that competition must be based on direct confrontations. Examination of numerous adults of all five species has revealed signs of quite serious intraspecific fighting: tattered ears, broken or amputated tails, fractured digits, damaged eyes, body scars, etc. The source of these wounds can be reliably traced to intraspecific fights since not one of 147 prepubertal individuals of all five species examined during the study exhibited such damage. Among adults examined, healed wounds were found with 13% of the *Galago demidovii* (N = 60), 20% of the *Galago alleni* (N = 20), 23% of the *Euoticus elegantulus* (N = 34), 22% of the *Arctocebus calabarensis* (N = 32). These data represent minimum figures, since certain old wounds may pass unnoticed. Traces of wounds are two to four times more common among males than among females, and this indicates that competition is probably more intense between males. However, actual fighting seems to occur only rarely under natural conditions in terms of its frequency in the behavioural repertoire. With the population of *Galago demidovii* followed in the forest, of 21 animals which were re-examined (11 after 1 year; 3 after 2 years; 2 after 3 years; 5 after 4 years) only one exhibited traces of recent wounding.

Given this infrequency of fighting it is virtually impossible to observe actual confrontations between individuals in the forest. For this reason, captured animals were artificially introduced into the home ranges of conspecifics in order to observe the effects. (These tests involved 7 *Perodicticus potto*, 3 *Galago demidovii* and 7 *Euoticus elegantulus*.) In almost all cases, the introduced animals eventually installed themselves in 'vacant' areas of the forest or disappeared, without necessarily engaging in a confrontation with the home-range occupants. (Males avoided the home ranges of other males, whilst females avoided home ranges of resident females.) On two occasions, individual *Euoticus elegantulus* females were observed being chased by the resident female in the area of release. The latter followed the intruders over some distance and wounded one of them fatally. On another occasion, a male potto introduced into the home range of another male led a nomadic existence in the surrounding forest before eventually returning to the resident male's range. The following briefly lists the events observed ($\male$ i = introduced, vagabond male; $\male$ 1 = resident male, 'controlling' the ranges of two females):

21.15 hrs: $\male$ i remains immobile in a tree with dense foliage. $\male$ 1 moves directly towards him from a point 25 metres away. $\male$ 1 seems to be very excited and droplets of urine are seen at the extremity of his penis. He waits for 5 minutes before attacking.
21.20 hrs: $\male$ 1 moves abruptly towards $\male$ i, whilst the latter remains in a

defensive posture (head retracted between the arms). ♂ 1 grasps ♂ i's back with one hand, pulling him forward and biting him. ♂ i bites his adversary's hands. After a bout of fighting lasting one minute, the two pottos move apart; ♂ i remains immobile whilst ♂ 1 circles around rapidly in the vicinity.

21.30 hrs: ♂ 1 returns to the attack and grasps ♂ i's back with all four extremities with ♂ i suspended from a small-calibre branch. ♂ i turns round and bites ♂ 1's left wrist, which bleeds. After 30 seconds, they separate again. ♂ 1 circles around ♂ i, with the latter remaining immobile as before.

21.40 hrs: ♂ 1 approaches to a point 60 cm away from his adversary and makes strikes with his teeth which fail to reach ♂ i but shake the foliage. ♂ i flees, followed by ♂ 1. ♂ 1 grasps ♂ i's fur with his hands and feet and is dragged a distance of one metre, biting him repeatedly before releasing him. ♂ i takes flight, at first moving quite slowly and then accelerating as he moves away from his adversary.

The next day, ♂ i was captured 500 metres away from the scene of the fight, having left the study area and crossed a roadway.

During the time of the year when this fight between the two pottos was observed, the two females associated with ♂ 1 (♀ 4 and ♀ 2) were both suckling infants and were sexually inactive. Thus, this fight obviously represented territorial defence independent of the period of mating.

The term 'territory' has been strictly defined for use in ethological studies. A 'territory' is defined as *an area occupied by an animal, or group of animals, and actively defended against conspecifics* (see Richard, 1970). With many mammal species it is extremely rare to observe actual fighting under natural conditions, and if the definition of a territory is followed to the letter one is practically obliged to abandon the use of this term altogether. This is why the term 'home range' has been used in most places throughout this account of prosimian behaviour, involving a much less restrictive definition: *The home range is the area covered by an animal, or group of animals, in the course of its movements.* Nevertheless, in captivity the five lorisid species all exhibit aggressive behaviour towards their conspecifics (Charles-Dominique, 1974a), this aggressivity being exacerbated by permanent contact between individuals. Under natural conditions, territorial equilibrium is attained with a minimum of fighting and various mechanisms intervene to bypass direct confrontations which might be prejudicial to the survival of the species.

The first mechanism can be simply termed *avoidance*. Any animal penetrating into the home range of a neighbour is made aware of this by olfactory marking sites and will leave the area quite rapidly. When individuals are maintained together in a common cage, on the other hand, they are unable to avoid one another and to maintain isolation. They rapidly become subject to stress, and in many cases an individual may eventually succumb from the effects. This is what usually occurs when too many individuals of a given lorisid species are maintained in one cage, even if no actual fighting occurs. Sometimes, avoidance is not achieved even under natural conditions. With bushbabies, in such cases, before engaging in a full attack the home range occupant may intimidate the intruder by leaping around vigorously nearby. This behaviour often suffices to bring about retreat by the intruder, involving a downward plunge which may even take the retreating animal to ground level. This has been seen on two occasions with needle-clawed bushbabies in the field, and it has frequently been seen in captivity with the other two *Galago* species.

When fighting occurs between bushbabies, the eyes are partially closed and the ears are fully retracted, thus reducing the risk of injury. The hands are always used to grasp the opponent's fur prior to biting. It was observed with fights between Demidoff's bushbabies in captivity that if one of the combatants uttered a distress-call, the adversary would at once release its grip. This mechanism permits termination of the fight once one protagonist has begun to gain the upper hand.

(iii) *Male-female relationships*

Male-female relationships are not strictly limited to actual mating; they persist throughout the year although females of all five prosimian species usually have only one infant a year. Contacts between the sexes usually occur during the night-time activity phase, however, since the males almost always sleep separately from the females.

1. Courtship behaviour

As a rule, one tends to associate courtship with the act of copulation, since these two components of behaviour are closely linked in numerous groups of animals. But among the lorisids courtship serves to establish social bonds between individual males and females which will not in fact mate with one another until much later (often several months after the female has accepted the male). Any new female not 'controlled' by another male is courted, whether she is prior to puberty, lactating, gestating or at any other phase of the sexual cycle. Courtship is always initiated by chasing of the female by the male.

(a) *Galago demidovii*. In this species, some males (8 observed cases in captivity) attempt to approach the female very cautiously, behaving as a 'subordinate' towards her. When the female permits the male to lick her face, the male extends his licking actions further and further until the female breaks away and utters aggressive calls. The male's approaches are repeated and eventually the female will respond by licking the male in her turn. When the male is thus 'accepted', allogrooming between the two animals may last for several hours. In the course of such grooming, almost all parts of the body are subjected to licking. When engaged in allogrooming, the two partners will often hang suspended beneath a branch by their hind-limbs, exhibiting very intense reciprocal grooming interrupted from time to time by bouts of 'play behaviour' in which one animal will hang suspended from the other.

In all cases where such courtship was observed, the males concerned were of small or medium size. In three other cases in captivity, involving quite large *Galago demidovii* males, much more intense chasing behaviour was observed, with the male attempting to grasp the female by force. Frequently, the male was seen to catch the female 'in mid-air', subsequently remaining suspended by his hind-limbs from the branch with the female held by a limb or by the tail. Whenever a female in this position turned around to bite her captor, she was immediately released. Initially, such chasing frightens the female and in captivity she is unable to move sufficiently far away from the male. Progressively, however, the male's chasing becomes less intensive and the female permits him to approach closer and closer until the two partners eventually engage in reciprocal licking behaviour.

During courtship the male frequently marks by depositing urine or by rubbing his anogenital region on supports. He will also rub his cheeks in the female's urine-marks. When the female has 'accepted' the male, she also exhibits urine-marking. Thus, in the course of the subsequent bouts of allogrooming, the urine borne on the hands of the partner must surely contribute to reciprocal marking. Quite often the male will utter the squeak – or, more rarely, the 'tsic-call' – and frequent visits to the female's nest are made. However, there are marked individual variations in the frequencies of these vocalisations.

Following acceptance of the male by the female, the period of intense allogrooming continues for several days and then the partners progressively 'calm down'., Soon afterwards, licking bouts between them reach a maintenance level of 30 seconds to 15 minutes (varying according to the individual involved), as opposed to the bouts of one to two hours observed at the outset. However, after a period of separation lasting several days allogrooming lasts longer than usual. The social bond established between the male and the female seems to be persistent since even after separation for a period of three months in captivity a male and female linked in this way will recognise one another immediately when contact is re-established, and they immediately engage in intensive allogrooming.

(b) *Euoticus elegantulus.* On three occasions the author has observed the courtship behaviour of the needle-clawed bushbaby in captivity. In every case the male pursued the female and attempted to capture her by force, though without any attempts to bite her. After several days had elapsed, the male's chasing behaviour became less intense and the female permitted him to approach closer and closer to the point where allogrooming could occur. As with *Galago demidovii*, such initial allogrooming sequences may last one hour or more.

(c) *Perodicticus potto.* The courtship behaviour of the potto can be more easily observed in the forest because of the slow locomotion of this species. On two separate occasions the author was able to observe the courtship behaviour of free-ranging male pottos, directed towards females released in the forest near the laboratory.

♀ 4 and ♂ 1. ♀ 4 was a female accompanied by a suckling infant, released close to the laboratory on 13 January 1967. ♂ 1, a resident male already 'controlling' another female (♀ 2) came to visit the new female for the first time on 2 February 1967. On almost every subsequent night, this male was sighted at about 21.00-22.00 hrs entering the forest area occupied by ♀ 4 and proceeding to follow her for one or two hours. The female initially fled, often descending to the ground and covering 20-50 metres in a straight line to climb up another tree. No direct contact was observed, but at the end of March 1967 the two animals were often sighted within a dozen metres of one another with no apparent attempt by the female to flee from the male. On 22 March (48 days after the initiation of courtship) the female was observed following the male for part of the night. It is possible that direct contacts did occur between them, but escaped observation.

♀ 3 and ♂ 6. ♀ 3 was an immature female approximately one year old. She was released under the same conditions as ♀ 4 on 31 December 1966, but she remained in an 'island' of trees isolated from the rest of the forest. It was not until 15 February 1967 that ♂ 6 (a vagabond male introduced into the area one month previously) discovered the presence of ♀ 3. Thereafter, he came to visit her regularly, each night crossing a deforested zone of 30 metres to reach her. After four nights of chasing by the male, the female accepted the presence of the male nearby, but threatened him whenever he approached too closely. Seven days after the initiation of courtship, she permitted the male to lick her, and he approached whilst uttering 'tsic-calls' (type A calls) to lick her face, neck and shoulders. Whenever he extended the range of licking too far along her back, or towards her genital region, or attempted to mount her, the female repelled him with aggressive calls. On such occasions, the male moved away to a distance of about one metre and waited immobile, sometimes for as long as 15 minutes, before returning to resume his licking of the female. Several times, the female broke away from the male to hang suspended from the branch by her hind-limbs. The male then followed suit and continued to lick her in a manner reminiscent of that already described for the courtship of *Galago demidovii* (above). The active licking phase lasted three days (22, 23 and 24 February), after which time the relationship between the two animals became more discreet. No further allogrooming was observed, but the male and female were often sighted only 10-20 metres away from one another during the night. During the daytime, however, they always slept separately at a distance of more than 100 metres away from one another. (At this point, the female was captured in order to check her sexual condition. The vulva was sealed off and the genital organs had not yet reached adult development.)

It should be noted that two recent studies of the potto in captivity have demonstrated that there are exchanges of odour which occur in the course of allogrooming, involving scrotal or 'labial' glands (Manley, 1974) and reciprocal urine-marking (Epps, 1974). On the other hand, the author has never observed the utilisation of the apophyses of the cervical vertebrae in contact behaviour between males and females, of the type suggested by Walker (1970) and Epps (1974).

As a general rule, courtship corresponds to the establishment of bonds between a male and a female, independent of the age or sexual condition of the latter. When the male is accepted by the female there is an intensive phase of allogrooming which persists for several days. Thereafter, the two animals become more independent with respect to one another, but they are nonetheless united by a persistent social bond.

During the period following courtship, direct contacts are brief and discreet in the galagines and virtually non-existent in the potto. The male is nevertheless far from disinterested in his female(s), but he limits his interest to rapid (direct or indirect) almost nightly investigations of the sexual state of his partner(s). This may be referred to as 'male visiting behaviour'.

2. *Male visiting behaviour*
(a) *Galago demidovii*. It is almost impossible to follow the

behaviour of a male *Galago demidovii* in the forest in a continuous fashion. However, it has happened that the same 'Central A' male has been captured at separate points, in the home ranges of different females, in the course of one night.

In captivity, an attempt has been made to recreate the natural conditions encountered by Demidoff's bushbabies. Three adjacent cages with communicating passageways are used to house a male and his two females (already subjected to courtship), with all three animals sleeping separately. At the onset of nocturnal activity, the male is allowed access to the first female. He immediately pursues her, uttering the 'rolling call' (type B call: contact-seeking). The female avoids him for 15-30 seconds and then permits him to approach her. Thereupon, the male sniffs at her genital region, frequently exhibiting urine-marking, and the two partners engage in mutual licking. They then separate rapidly, and the male attempts to pass along the passageway leading to the second female. As soon as the passageway is opened, he exhibits exactly the same behaviour as with the first female.

This experiment has been repeated with seven different trios of Demidoff's bushbabies and the behaviour of the male in each case has been found to follow exactly the same pattern: the female is visited, her genital region is subjected to olfactory investigation, urine-marking occurs, followed by allogrooming; then there is a period of independence with respect to this first female, followed by seeking out of the second female which has not yet been visited. The only elements subject to individual variation are the intensity of following of the female prior to contact, the frequency of urine-marking and the duration of allogrooming.

(b) *Galago alleni*. The only information as yet collected derives from a male (♂ 14) which was equipped with a radio-transmitter in the field between 27 June and 17 July 1973. During this period, five females all 'controlled' by this one male (♀ 5, ♀ 6, ♀ 8, ♀ 17, ♀ 18) were also carrying radio-transmitters broadcasting at different frequencies, which permitted individual identification of these females and precise pinpointing of their location in the forest. However, at least three other females controlled by this male, further to the west and to the north (see Fig.58), were not captured. The male involved spent each day sleeping within a restricted zone of his home range, and went around to visit the females during the night:

27.6.73: visits to home ranges of ♀ 17, ♀ 5, ♀ 6, followed by ♀ 8 and ♀ 18.

28.6.73: visits to home ranges of ♀ 5 and ♀ 6. Allogrooming observed.

29.6.73: male moved off to the west.

30.6.73:⎫ male remained every night in the home range of
1.7.73:⎬ ♀ 8 and ♀ 18. Allogrooming seen with ♀ 8. ♀ 8
2.7.73:⎭ captured and found to be in oestrus.

3.7.73: male visited home ranges of ♀ 8, ♀ 18 and ♀ 17.

4.7.73:⎫ male remained for nine nights with ♀ 17, the latter
to ⎬ coming into oestrus on 8.7.73 (examined on
12.7.73:⎭ capture).

13.7.73: male moved off to the north.

14.7.73: male visited home ranges of ♀ 5, ♀ 6 and ♀ 17.

15.7.73:⎫ male moved off to the north.
16.7.73:⎭

17.7.73: male visited home ranges of ♀ 5 and ♀ 6.

Allogrooming between the male and these females was only observed on two occasions, but the observations were discontinuous. As with Demidoff's bushbaby, meeting between the male and the females must be relatively brief, but the male remains within the range of each female for quite some time. Several females may be visited in the course of a single night, though when one of them comes into oestrus the male will not leave her for a number of days and will consequently fail to visit the other females. In the case of ♀ 5 and ♀ 6, the male remained for a period of up to 16 days.

(c) *Perodicticus potto*. The visiting behaviour of the male potto proved to be amenable to study in the forest near to the laboratory at Makokou. In 1967 and 1968, the various sectors of the forest were already linked to one another by lianes, as has already been explained in connection with the relationships between neighbouring pottos (see Figs. 65, 66). The northern part of the study area was occupied by one female (♀ 2), whilst the southern part was occupied by another (♀ 4). Both of them were associated with the same male (♂ 1), which spent most time in the northern area but came to visit ♀ 4 regularly. In order to visit this female, he passed across lianes on which electrical contacts had been attached (Fig.75) and linked to a graphic recorder. Of course, information obtained by this indirect process was supplemented by direct observation.

Following the courtship period, during which daily visits occurred, the male's visits became less frequent, occurring every 2-4 days (Fig.76) and lasting between half-an-hour and three hours. The male and the female apparently did not encounter one another directly, but ♂ 1 very often passed close by the artificial food-source where ♀ 4 came to eat almost every night. The male only rarely ate food there, but he was observed on three occasions to engage in marking behaviour. The female also marked with urine near to the artificial food-source, and she was once discovered marking at the same spot marked by the male the night before. (The food-source, which was supplied with bananas every day, was placed at a height of one-and-a-half metres in a small tree, thus facilitating observations.) The visiting behaviour of the male apparently occurred throughout the year. It was observed from March to May 1967; the author was then absent from Gabon from June to September, but visiting behaviour was again seen regularly from October 1967 (when observations were resumed) to 19 March 1968 (the day when ♂ 1 disappeared). Almost all of the visits recorded occurred at approximately the same time. For example, in February 1968 the following times were noted: 21.50 hrs, 22.20 hrs, 22.30 hrs, 22.15 hrs, 1.15 hrs, 22.15 hrs, 22.02 hrs. Visits always took place at least two and a half hours after the pottos had become active at night, and ♂ 1 always engaged in feeding before visiting ♀ 4.

In the case of these two pottos it would seem that the male and the female communicated principally by means of olfactory signals deposited by one animal and investigated by the other some time later. This system of 'deferred communication' can be interpreted as an adaptation to the locomotor characteristics of members of the subfamily Lorisinae. The slowness and discreetness of movement would not permit the partners to meet regularly in the forest, as

Figure 76. Visits made by potto ♂ 1 to potto ♀ 4 in the period March-May 1967. In order to visit this female, the male made use of a liane bearing an electrical contact.

happens with the bushbabies. However, it would seem that the relationships between ♀ 9 and ♂ 10 (monogamous male) in 1969 were slightly less indirect than those found in the case discussed above. In November and December, these two pottos were twice seen arriving together at an artificial food-source (at 22.30 hrs and 1.30 hrs, respectively), with the male following the female at a distance of 2-20 metres. Yet no actual physical contact was observed.

3. *Mating*

(a) *Galagines.* The great mobility of bushbabies within their home ranges permits males to encounter females in oestrus. The example outlined above for *Galago alleni* shows that once a male has located one of his females just prior to oestrus he will stay close to her for several days. In captivity it has been noticed with *Galago demidovii* that there is an increase in the interest a male exhibits in a given female when she is close to oestrus. One observes chases, frequent olfactory investigation of the female's genital region and mounting attempts, which are rejected by the female. Such behaviour is first observed 2-5 days before full opening of the vulva. Copulation always takes place in the same manner: the male straddles the female's back, grasping her heels with his feet and clinging to her flanks with his hands (typical primate mounting posture).

With *Galago demidovii*, 25 copulation sequences have been observed in captivity. In all cases, pelvic thrusting lasted 15-20 seconds and was followed by spasmodic movements of the tail, which was swung sharply from side to side (movements probably associated with ejaculation). The glans penis of bushbabies is covered with cornified spicules which maintain anchorage on the vaginal wall. When the male has terminated, he will attempt to separate from the female, but sometimes he is unable to do so. When this happens, the female starts to utter weak vocalisations and attempts to bite the male, whilst the latter keeps clear by moving his head from one side to the other and extending his arms. The duration of this post-coital attachment usually varies from 45 seconds to a few minutes, but in exceptional cases may be as much as one hour. Some pairs only copulate once or twice, but in the author's captive colony there is one male (a dominant animal of large body-size) which mates throughout the night when a female is in oestrus. Copulation may generally occur during two successive nights, but never more than that.

During copulation the female is always suspended beneath a branch, maintaining a grip with all four limbs. In fact, a male attempting to copulate will adopt a position beneath the branch where a female is suspended, or may sometimes pull the female with one hand in order to swing her beneath the branch, so that a final suspended position is always adopted. Normally, the female will spontaneously adopt a position beneath the branch when she is ready to copulate.

In at least two other *Galago* species (*Galago senegalensis* and *G. crassicaudatus*: Doyle, et al. 1967; Manley, pers. comm.), the female remains above the branch for copulation. The same would seem to be true of *Euoticus elegantulus* (one observation in the forest: Dubost, pers. comm.). The inverted position adopted by *Galago demidovii* is doubtless related to the habitat of this species, which is dominated by fine and relatively flexible branches and lianes.

(b) *Lorisines.* The slow locomotion of these prosimians does not permit frequent encounters of the kind found with the bushbabies, but 'deferred communication' by means of urine-marking permits the male to recognise the approach of oestrus in the female. Experimentally, Seitz (1969) has demonstrated that in the potto there is a resurgence of the male's interest in the urine of a female in oestrus. In fact, in the forest the author has observed behaviour of a male potto which similarly indicated perception of the approach of oestrus (observation of ♂ 1 and ♀ 4 in 1967, Fig.76):

Whereas, for a period of several months, ♂ 1 had been visiting the home range of ♀ 4 once every 3-4 days, the frequency of visits increased from 9 May onwards.

As from the 21 May, ♂ 1 came to visit the home range of ♀ 4 regularly every night just after nightfall (at about 19.00 hrs), whilst his previous visits usually occurred at about 21.30 hrs.

On 27 May, ♂ 1 slept 25 metres away from ♀ 4.

On 2 June, ♀ 4 followed the male for part of the night.

On 5 June, ♀ 4 was captured and vulval examination showed the concluding phase of oestrus, which probably reached its peak on about 2 June. Thereafter, the male's visits became less frequent again.

On 22 November, 1967, ♀ 4 gave birth to an infant.

In the light of these observations, it would seem that the visiting behaviour of the male potto was slightly modified 3 weeks before the female's oestrus and markedly modified 12 days beforehand.

(iv) *Mother-infant (and infant-adult) relationships*

1. *Galagines*

The stage of development attained by bushbaby infants at the time of birth is much less advanced than that found with lorisine infants. This is perhaps related to the fact that the bushbabies are typically nest-living.[1] Some species build leaf-nests (Fig.77), as is the case with *Galago demidovii* (Vincent, 1968; Charles-Dominique, 1972), *G. senegalensis* (Sauer & Sauer, 1963) and *G. crassicaudatus* (Bearder & Doyle, 1974). Others make use of tree-holes, which may be lined with a few leaves. *Galago senegalensis* may also use nests of this kind, whilst *G. alleni* is exclusively found in tree-hollow nests. This same behaviour is found with certain Malagasy lemurs, where nests may also consist of leaves or branches (*Microcebus murinus*, Petter, 1962; Martin, 1972a; *Microcebus coquereli*, Petter et al., 1971; *Daubentonia madagascariensis*, Petter, 1962), or be represented by tree-hollows (*Microcebus murinus*, Martin, 1972a; *Cheirogaleus major*, Petter, 1962; *Phaner furcifer*, Petter et al., 1971; *Lepilemur mustelinus*, Petter, 1962).

With Demidoff's bushbaby, a female will isolate herself

---

[1] *Euoticus elegantulus* has in fact never exhibited nest-utilisation in captivity, but nothing is known about the state of the infants at birth in this species.

Figure 77. Nest of Demidoff's bushbabies. (Photo: A.R. Devez. Reproduced from Charles-Dominique, 1972 with the kind permission of Paul Parey Verlag)

from the matriarchal group before giving birth and will only re-join the group one to two weeks later. The neonate is covered with a sparse coat of hair which leaves the skin exposed to view. The eyes are only partially opened, and only the hands are capable of grasping an object. For example, if the neonate (Fig. 78) is suspended from a branch, it will hang from its hands alone. Nevertheless, the infant is able to crawl into the nest and snuggle underneath the mother to reach her teats.

There is intensive contact between the mother and her infant. At birth, she licks and manipulates the neonate and devours the foetal membranes. Her responses to other strange bushbabies or any predators become much more aggressive at this time, and she frequently utters alarm-calls. However, in captivity any closely associated bushbabies (the male and

A

Figure 78 (A above, B overleaf). Neonate *Galago demidovii*. (Photos: A.R. Devez. Reproduced from Charles-Dominique, 1972 with the kind permission of Paul Parey Verlag)

B

young from previous litters) are accepted in the nest only a few hours after the birth, and these conspecifics eagerly come to lick the neonate. At the slightest sign of danger, the mother grasps the infant in her mouth, taking a hold on the flank fur, and flees. Under natural conditions, any *Galago demidovii* nest which is disturbed is at once abandoned.

Very soon afterwards, the nest is utilised only for sleeping during the daytime. From three to seven days after birth the mother begins to carry her infant out of the nest at dusk and to 'park' it, in the vegetation, where it will remain immobile. (This has been observed with both *Galago demidovii* and *Galago alleni*.) Whilst the baby is 'parked' in this fashion, the mother feeds in the vicinity. If she moves on to another feeding area, she comes to collect the infant and then re-deposits it in the vegetation further on. As dawn arrives, she will collect the infant for the last time and carry it back to the nest with her. When coming to collect the infant, the mother calls with a 'rolling call' (*Galago demidovii*), a 'croak' (*Galago alleni*) or a 'tsic-call' (*Euoticus elegantulus*), and the infant responds with high-pitched vocalisations ('click' and 'tsic' – see section on vocal communication). The infant will only call as a response to its mother or, as an exception, spontaneously if it is isolated when dawn is approaching.

Whilst clinging immobile in the foliage, the young

bushbaby is practically immune from discovery and it can only be located by means of eye reflections produced with a headlamp. If approached by an observer, the infant will normally remain completely still; but as soon as the branch is touched it will allow itself to fall. Once it has fallen into dead leaf litter, it will again become immobile and hence extremely difficult to find. Sometimes, in the course of its fall, the infant will utter a distress-call which induces the mother to appear rapidly on the scene in the space of 15-20 seconds. If the observer remains within a radius of approximately 1 metre of the point where the infant has fallen, the mother will approach whilst producing alarm calls. If, on the other hand, the observer moves away a distance of several metres, she will retrieve her infant and flee.

At the age of two to three weeks, the young bushbaby begins to move around independently and actively and will wander around in the vicinity of the nest or close to the site of 'parking' by the mother. A dense covering of fur appears at this same stage of the infant's development.

At one month of age, the young Demidoff's bushbaby will begin to eat with its mother, snatching fragments of food (e.g. insect legs) directly from her mouth. In this particular species, weaning occurs at about the 45th day of life. The growth-curve tends to flatten out at this period of development, usually with complete arrest of weight-increase; but at about the age of 2 months, body-weight starts to increase again.

The definitive body-weight is attained at 5-6 months of age in *G. demidovii*, at 6-8 months in *G. alleni* (see Fig.79), and at 8-10 months in *Euoticus elegantulus*.

As the young bushbaby gradually develops, it begins to follow the mother more and more during her nightly excursions. In so doing, it is exposed to progressive dietary conditioning. Sometimes, when the growing infant is unable to leap across a gap after its mother it will utter a call. In such cases, the mother returns, grasps her offspring in her mouth, leaps across the gap again and deposits the young animal on the other side. This behaviour has been observed by the author under natural conditions with *Galago demidovii* (with infants of about 1 month of age), with *Galago alleni* (with infants aged 45 days) and with *Euoticus elegantulus* (with infants 1-2 months old). With the needle-clawed bushbaby, the

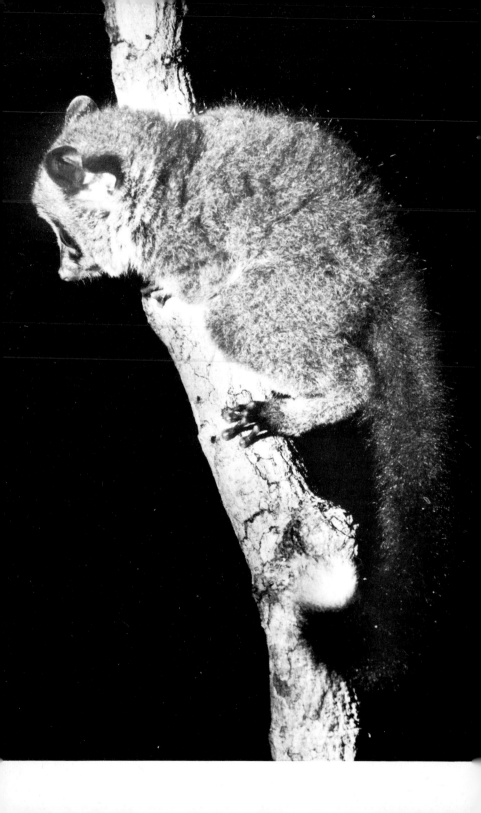

mother may grasp her infant by its flank in her teeth (as with the other two bushbaby species), but sometimes the young animal will climb on to her back and grasp her dorsal fur. When the infant rides on the mother in this way, she will jump as usual. However, as she arrives in the foliage at the landing-point the infant will release its hold on the mother and independently take a grip on the foliage.

During nocturnal 'siestas', the mother and her infant engage in prolonged bouts of grooming interspersed with play behaviour in which the young bushbaby leaps on the mother, nibbles her, hangs from her tail, bounces up and down on the spot, and so on. When there is more than one infant (twins or infants of different females associated in a 'matriarchy'), there may be group play.

A young female will follow her mother for a considerable period of time. Even when she has reached adulthood, she may follow her mother for part of the night, at a distance of 10-20 metres. At least, this has been observed with *Galago demidovii* under natural conditions. As a general rule it is quite a common occurrence with all three bushbaby species to trap two females in close proximity to one another during the night and to find that one is a multiparous adult (with developed mammae), whilst the other is a nulliparous adult (with undeveloped mammae). With *Galago demidovii*, of 5 females first recorded as immature animals in their mothers' home ranges, 3 were later recaptured at the same sites after intervals of 2, 4 and 5 years, respectively. Similarly, with *Galago alleni* (studied more recently) 2 out of 3 females first recorded as immature specimens were later found after intervals of 1 and 2 years (respectively) in the same areas. If allowance is made for mortality under natural conditions, it seems likely that young female bushbabies generally remain sedentary. This would form a natural basis for the creation of groups of adult females in 'matriarchies', with the females in each group regrouping at dawn every day to return to the sleeping-site, together with any infants. (The social relationships between the members of such groups will be examined in the following chapter.)

The young males also follow their mothers as soon as they are able to do so. The author had the opportunity of studying

Figure 79 (left). Young Allen's bushbaby at about 2 months of age. (Photo: A.R. Devez)

several young male *Galago demidovii* in the forest. They were found to follow their mothers, but they were also found to follow other females belonging to the same social group ('matriarchy'). At about the age of 5-7 months, each of these young males was also seen to follow the Central A male (dominant male) associated with his mother. In this way, the young male's home range comes to coincide with that of the Central A male, and the latter is followed into the home ranges of other females with which he is associated. However, the young male never enters the home ranges of females 'controlled' by any other Central A male. On 12 separate occasions, single young subadult males born to one of the associated females have been captured in the same trap as a Central A male (observations involving 4 different Central A males and 5 different young males). Two bushbabies can only be trapped together in this way if they actually enter the trap together: consumption of the food bait requires no more than 30 seconds to 1 minute, and any bushbaby attempting to leave the trap will release the closure mechanism.

These young males retain such social relationships with their group for some time, re-uniting with the females and very young animals at dawn in order to move on to the sleeping-site. At about the age of 8-10 months, when puberty is reached, the young males eventually leave the area to adopt a vagabond or nomadic existence. Of 7 young male *Galago demidovii* first recorded with their mothers in the study area between 1968 and 1972, not one was recaptured after the age of 10 months.

2. *Lorisines*

At birth, infant lorisines have their eyes completely open and already possess a dense covering of fur. With both the potto and the angwantibo, the mother devours the foetal membranes and cleans the neonate (as in the galagines), but the infant is never manipulated. Parturition takes place on a branch and if the infant is too weak to grasp the mother (as occurs on occasion in captivity when birth takes place prematurely), it will fall to the ground without any attempt by the mother to retrieve it. Under typical conditions, however,

Figure 80 (right). Angwantibo mother returning to collect her neonate, which has been left suspended from a branch.

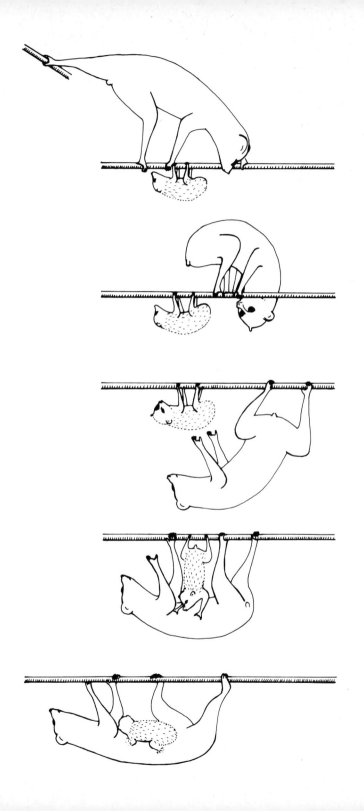

the neonate will climb slowly in its mother's fur, seeking to take up a position on her belly. The author once observed (Charles-Dominique, 1966a) the birth of a baby angwantibo: the infant moved slowly over the mother's right flank, climbed on to her back, continued on to her left flank and then passed between the two limbs on that side to adopt eventually a longitudinal orientation on her belly. This entire process took 1 hour. In the course of the very first night following birth, the mother moved off in search of food, with the infant clinging to her belly. In the course of resting periods during the night, however, she assiduously licked the baby, which made no attempt to suckle. Suckling always takes place during the daytime, whilst the mother is sleeping.

From 3-8 days after birth onwards, the mother begins to 'park' her baby at dusk and to retrieve it as dawn approaches (Fig.80). This behaviour has been repeatedly observed both in the forest and in the laboratory with angwantibos and pottos alike. Deposition of the infant always takes place in the same fashion: in the course of the bout of grooming immediately following awakening at the beginning of the night, the mother spends a considerable time licking her infant, which becomes excited, first moving around on her fur and then eventually taking hold on a branch. The mother then continues to lick her offspring before setting off alone. Thereafter, the very young angwantibo remains suspended from the branch. However, deposition of the infant is not entirely regular; a mother potto may leave her baby clinging to a branch throughout one night and yet keep it clasped to her belly the next night. Nevertheless, this 'parking behaviour' is a typically recurring phenomenon in these two lorisine species and it is very reminiscent of 'parking' of young bushbabies outside the nest at night.

The mother returns to retrieve her infant during the early morning. With both pottos and angwantibos, the mother utters a high-pitched 'tsic-call' which evokes a similar vocalisation from the infant, thus permitting its localisation in the vegetation. During the initial stages of such parking behaviour, the mother angwantibo retrieves her infant by arranging her body as a sling beneath the branch, hence encouraging her offspring to take a grip on her fur (Fig.80). Later on, she simply moves alongside the 'parked' infant,

Figure 81. Neonate potto.

which independently approaches to cling to her belly. With the potto (Figs. 81, 82), on some occasions a mother has been observed to assist her infant to climb on to her by gently pushing it towards her belly with one hand. At a still later stage, with both pottos and angwantibos the mother and her infant move towards one another whilst communicating with 'tsic'-calls.

As the young potto or angwantibo develops, it begins to move around the site where it has been 'parked' by the mother. In some cases, an infant of either species may fail to leave its mother at the beginning of the night and will allow itself to be carried around, clinging to her dorsal fur (Figs. 83, 84.) Such carriage does not continue throughout the night, however, for the infant occasionally descends from the mother and starts to follow her, until at some point she happens to move too far ahead and then continues to move away without responding to her offspring's calls. Subsequently, she will only return to retrieve the infant as dawn approaches.

Figure 82. Potto mother carrying her infant on her abdomen. (Photo: A.R. Devez)

Soon afterwards (at the age of 3-4 months in the potto), the young lorisine begins actively to follow its mother virtually everywhere, either riding on her back or walking just behind. Thus, dietary conditioning commences (see p.49) at the same time that weaning begins to take place. At about 6 months of age, the young potto moves around alone, no longer sleeping with its mother; but it may meet her from time to time in trees which are in fruit. In this way, mother-infant contacts become progressively less common.

In the study area in Gabon, it was only once possible to observe the weaning of a young (female) potto taking place under natural conditions. At 6-8 months of age, the young female gradually slept with her mother less frequently, and by 8 months she was sleeping alone. However, she continued to share her mother's home range, and the two animals occasionally met at the artificial food-source or in a tree in fruit. 10 months after birth, the mother moved on to establish herself in a new home range 200 metres away, and she was still recorded in this new range two years later. Her daughter, on the other hand, remained in her mother's previous home range, but was accidentally killed at the age of 16 months by an electric cable.

Of three young male pottos born in the study area whilst observations were in progress, two moved on and one remained in the area of its birth. The latter case was an exception in that the only adult male present disappeared when the young male was only four months old. Hence, the young male remained within his mother's home range and at the age of one year, whilst still prior to puberty, he began to visit a second female in the same manner as an adult male. It should be noted that the study area was connected to the main block of the forest only by a narrow strip of trees 200 metres in width lining the bank of a river, and this doubtless placed a limit on the passage of vagabond males (only two sighted in the course of three years). It seems likely that in the heart of

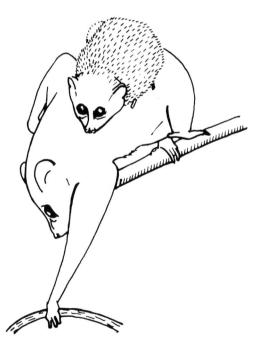

Figure 83. Young angwantibo riding on its mother's back. (Drawing prepared from an original photograph)

Figure 84. Young angwantibo at the age of 2 months, still covered with its juvenile coat. (Photo: A.R. Devez)

the forest males which disappear are rapidly replaced. Accordingly, it is likely that this young male would have departed like the other two if an adult male had been present.

In fact, it would seem that the departure of a young male potto from its mother's home range is generally related to the presence of an adult male. However, the author has never observed – either in the field or in captivity – any aggressive behaviour directed towards young animals by the adult male or female at the time of weaning.

The following record was made of the departure of the young ♂ 5 at the age of 6 months. On 20.3.67, ♂ 5 left its mother's home range. On 21.3.67, the young male was spotted 300 metres away. On 25.3.67, he returned to the vicinity of his mother, remaining immobile for several hours at a distance of 20 metres from her. Subsequently, ♂ 5 moved off again and was never spotted in this area of the forest thereafter. The male associated with the young male's mother seemed to be 'indifferent' to the presence of the youngster.

(v) *Relationships within social groups ('matriarchies')*

Whereas with the lorisines adult females become progressively independent with respect to their daughters, the galagines are unusual among the lorisids in that certain adult females will associate in groups. This is not found only with the three Gabonese species but also with *Galago senegalensis* (Sauer & Sauer, 1963; Haddow & Ellice, 1964) and *Galago crassicaudatus* (Bearder & Doyle, 1974) in East and South Africa. Among the other nocturnal primate species, *Microcebus murinus* in Madagascar (Petter, 1962; Martin, 1972a, 1973) exhibits the same peculiarities: social grouping of adult females and their infants.

Females whose home ranges overlap partially or completely do not necessarily reassemble every morning to return together to a common sleeping-site. For example, with *Galago alleni* (Fig.64), it was found that ♀ 5 and ♀ 6, whose home ranges overlapped by 60%, only slept together on 11 occasions out of 21 nightly records whilst they were both carrying radio-transmitters.

This variability in reassembly of groups before dawn is also noted when a given sleeping-site is observed regularly night after night. For example, Struhsaker (1969) noted with *Galago demidovii* the following numbers of animals grouped together in one particular nest on successive nights: 2, 2 (or 4), 2, 3, 3, 2.

The author himself has noted with *Euoticus elegantulus* variation in size of a group reassembling in one given area before dawn between 2 and 7 (groupings of 2, 4, 5 and 7 individuals, respectively).

1. *Galago demidovii*
The following groups were found in leaf-nests (Fig.77):
> 5 groupings of 2 individuals
> 18 groupings of 3 individuals
> 4 groupings of 4 individuals
> 3 groupings of 6 individuals
> 1 grouping of 10 individuals

In contact groups merely huddled together in the foliage, on the other hand, the following groupings were found:
> 4 groupings of 4 individuals
> 1 grouping of 5 individuals
> 1 grouping of 6 individuals
> 1 grouping of (approximately) 10 individuals

Isolated individuals were only found on two occasions sleeping in foliage and on one occasion in a leaf-nest.

Information on the actual composition of these sleeping groups is fragmentary, since some individuals usually manage to escape when a nest is examined. With groups found sleeping in nests, individuals of identified age and sex turned out to be 11 adult females, 1 adult male and 11 subadults or infants. On the other hand, the two individuals captured alone in foliage during the daytime were both adult males. In any case, it is certain that females may occur together in the same nest.

2. *Galago alleni*
Tree-hollow nests were found to contain 42 isolated individuals, 17 groups of 2, 6 groups of 3 and 1 group of 4. The two males followed by radio-tracking (Fig.64) always slept separately, even when ♀ 8 and then ♀ 17 came into oestrus. It was definitely observed that adult females may sleep together in groups of 2 (N = 18) or 3 (N = 6).

3. *Euoticus elegantulus*
Bushbabies of this species always sleep in 'contact groups' in clumps of foliage. Groups of 2 individuals were observed

twice; groups of 3, twice; groups of 4, 3 times; groups of 5, 4 times and a group of 7 on one occasion.

In the forest it proved possible to recognise reliably 2 adult females, 2 young animals and 3 adults of indeterminate sex in a group of 7 individuals assembled together to sleep during the daytime. On another occasion, 2 males were killed, by a hunter, in a group of 5 individuals (the other 3 escaping).

Within the Gabon study area, an adult male and female which were released and subsequently followed met up every morning to sleep together. A second pair was maintained in a cage within the home range of the free-ranging *Euoticus*. From time to time, the two free animals came to sleep on the cage, in contact with the two captive needle-clawed bushbabies, which would in such cases press themselves against the cage-wire accordingly. On three successive nights in between, it was also observed that a strange male similarly came to sleep in contact with the two captive individuals.

It would thus seem that with all three bushbaby species in Gabon adult females may form groups, together with their offspring, whilst the males may join up with such groups far less frequently (with the exception of *E. elegantulus*, where males join groups more commonly). The two bushbaby species occurring in East and South Africa (*G. senegalensis* and *G. crassicaudatus*) exhibit exactly the same grouping tendency. Bearder & Doyle (1974) report the following group sizes under natural conditions for the two species in South Africa:

Table 10 Frequencies of nesting group size for the two bushbaby species in South Africa (after Bearder and Doyle, 1974)

Number of individuals in nest	1	2	3	4	5	6
Galago senegalensis (%)	40	30	23	5	1	1
Galago crassicaudatus (%)	46	26	18	6	3	1

Sauer & Sauer (1963) have also reported sleeping groups of *Galago senegalensis* in South-West Africa. In East Africa, Haddow and Ellice (1964) found 42 sleeping groups of *G. senegalensis* under natural conditions, of which 10 had no adult male, 29 had one adult male and 3 had two adult males.

The fact that young males become vagabonds (nomadic) at puberty, while young females remain sedentary in their area of birth, together with the observation of polygamous associations exhibited by adult males, indicates that the basic structure of these bushbaby groups is matriarchal. That is to say, the structure is founded on females of the same origin, accompanied by their offspring, whilst males may join groups occasionally as an additional feature.

The attempt has been made in the author's laboratory to recreate artificially female groups of this kind by maintaining Demidoff's bushbabies in cages with numerous compartments communicating with one another by sliding doors. Of 5 females so far born and reared in this system in captivity, 4 have remained with their mothers, sleeping in the same nest, whilst the fifth became very aggressive towards her parents at the age of 11 months, so that she had to be separated.

On another tack, the author tried to unite in groups adult females which were previously unacquainted. Of 11 previously isolated adult females, 4 rejected any association at all, whilst the remaining 7 formed groups of 2, 2 and 3 respectively. Once grouped together, these latter females became aggressive towards any further strange females they encountered. Thus, an individual factor is involved in such groupings, in that some females are more aggressive than others. This aggressivity with respect to strangers is increased when a group has already been formed, and it may therefore act as a contributory factor in limiting the size of such female associations.

In a different experiment with Allen's bushbabies, 5 adult females were grouped together in the same cage in Gabon, including 2 mothers and their respective daughters (captured together in the same tree-hollow one year previously) and an additional adult female captured in a different part of the forest. After two years spent in captivity in this artificial group, the 5 females were released on an island of 60 hectares, already inhabited by wild *Galago alleni*. The 5 females then split up into three subunits of 2, 2 and 1, respectively. One pair was a mother with her daughter, the second pair was composed of the strange female and the other daughter, whilst the other mother became isolated.

These experiments thus demonstrate that there are no

obligatory social bonds between a mother and her daughter once the latter has reached adulthood. Apparently, it is not the link of actual parental relationship but the fact of proximity of two females which become accustomed to one another (usually related to birth, of course, under natural conditions) which is responsible for creation of the social bond. Given the sedentary tendency of young females, there is a strong probability that natural groupings will have a family (mother-daughter) basis, but it is nevertheless possible that two strange females may associate with one another under a certain set of circumstances, just as was the case under experimental conditions.

In sum, then, the distribution of female home ranges, with varying degrees of overlap, corresponds to a resultant of the following factors: sedentary tendency of young females; fecundity of adult females; mortality rate; degree of 'sociability' between adult females.

Whereas relationships between occupants of neighbouring ranges have only a 'territorial' significance in the absence of direct contacts between the animals concerned, social relationships existing within a matriarchal group are quite complex, in some respects reminiscent of those found in simian primates. However, the bonds uniting the partners of each group are little evident during the phase of activity (night-time); their main expression occurs during the resting phase (sleeping group). This daily periodicity in social manifestations is clearly evident from diagrams indicating the frequency of emission of vocalisations of the 'social' type in the course of the night. Contact calls are relatively numerous at dawn and dusk, but rare during the main part of the night. The awakening of bushbaby groups has been observed on numerous occasions in the forest. As the light-intensity decreases, the animals begin to stir, yawning and licking both themselves and their partners. With Demidoff's bushbaby, as soon as one individual leaves the sleeping-site, all of the others follow suit; but the tendency to aggregate persists for a number of minutes afterwards such that one can at this time observe groups of these bushbabies moving around in single file. However, dispersion occurs fairly rapidly thereafter, but even then the animals continue to remain in vocal contact for about an hour by means of short calls which elicit further calls in response (initiation of alarm calls composed of 2-4 units – Fig.71). With the needle-clawed bushbaby, awakening occurs in identical fashion, but dispersion of the group is a more drawn-out process. The animals leave the sleeping-site independently of one another, often at intervals of several minutes, such that there may be a total period of 10-15 minutes between the exits of the first and the last individuals. Sometimes, as with Demidoff's bushbaby, after dispersion of the group, the cohesive tendency may persist with one animal, which will follow another. Invariably, the followed animal will turn round and repel the conspecific with aggressive calls, whereas only a few minutes previously the two animals were still huddled closely together.

This intolerance of contact as soon as the bushbabies awake does not exclude all social relationships during the night, however, since – in addition to visits made by the male and relationships between adults and young – it has been observed that a young (but adult) female may follow her mother. Nonetheless, it would seem that relationships between adult females of a particular group are mainly limited to the period when the animals are resting. This is somewhat reminiscent of the social groups of gregarious primates in which the maximum frequency of social contacts is found during the 'siesta' and at the sleeping-site.

When returning to sleep, and particularly when awakening just prior to dispersion of the group, all bushbabies exhibit reciprocal licking. The close contact which they have during the daytime sleeping period also doubtless ensures mixture of their respective body odours, thus reinforcing social cohesion.

The observations conducted on *Galago demidovii* in captivity indicate that the initiative in choice of the nest-site and construction of the nest falls to the old female of a group. Younger females will follow her, but if they participate in nest-building it is only in the form of occasional transport of a leaf in the mouth, without actually incorporating it into the nest structure. There is thus a certain degree of dominance which is exerted at the time when the animals return to sleep. On one occasion, the author observed with the captive colony that the neonate of a 'subordinate' female was carried off by the 'dominant' female. The latter was neither lactating nor gestating, but she licked the neonate and carried it in her mouth as if it were her own infant. The real mother was in the same cage, but kept her distance and did not attempt to retrieve her infant. It is therefore understandable that under natural conditions females isolate themselves from their groups in order to give birth.

Apart from these exceptional cases of choice of the nest and 'priority' in the care of the neonate, the females in a given matriarchal group do not seem to exhibit any hierarchical behaviour of the kind described for many gregarious species in relation to feeding. In captivity, two females 'associated' in a group can drink milk together or even eat from the same fruit without aggressive interaction. With insects, on the other hand, the situation is different. All individuals present will rush forward, and it is usually the nearest animal which seizes the prey. The other individuals in the case then run after the captor and succeed in pulling away a few fragments of the insect. However, such situations do not occur under natural conditions.

The problem of hierarchical organisation must not, therefore, be envisaged in the manner usual for diurnal, gregarious primates, in terms of tests of direct competition (e.g. with respect to food). With solitary, nocturnal animals it is in relation to the size and location of individual home ranges that any expression of a hierarchy must be expected, if such exists, between the adult females of a given matriarchal group.

Conclusions

Following the detailed analysis of the behaviour and ecology of the five different lorisid species in Gabon, it is possible to examine the significance of the results in the broader framework of primate evolution. In particular, it should be valuable to reflect upon the distinctions between nocturnal and diurnal habits among the Primates in the light of information obtained from this study of the exclusively nocturnal lorisids.

If the tree-shrews (family Tupaiidae) are excluded from the order Primates, as suggested by a number of recent authors (e.g. Grassé, 1955; Romer, 1966; Martin, 1975), there remains a total of 188 living primate species (Napier and Napier, 1967; Petter, in prep.). Of these, 37 are prosimians (23 Madagascar lemurs, 11 lorisids and 3 tarsiers) and 151 are simians. Among the simians, 64 are New World platyrrhines (35 marmosets and tamarins, 29 cebid monkeys) and 87 are Old World catarrhines (75 cercopithecoid monkeys, 11 hominoids). None of the catarrhines is nocturnal, and only one of the platyrrhines exhibits nocturnal habits (the owl monkey, *Aotus*), whereas two thirds of the prosimians (27 out of 37) are exclusively nocturnal. There is thus a fairly clear distinction between simians and prosimians in terms of adaptations for nocturnal life.

In a recent publication (Charles-Dominique, 1975), the attempt has been made to interpret the evolution of the major orders of mammals in terms of differential ecological specialisation for diurnal or nocturnal life. In numerous biological disciplines little account is taken of this distinction in interpreting natural phenomena. Yet in a large number of cases taxonomic groupings correspond to clusters of animals sharing either diurnal or nocturnal habits. The insectivores are typically nocturnal, as are the bats, and nocturnal life is common among the rodents (with the exception of the squirrels) and among prosimian primates (Lorisidae, Cheirogaleinae, Lepilemuridae, Tarsiidae). By contrast, diurnal life is typical of the squirrels, the tree-shrews (with the exception of *Ptilocercus*), the majority of ungulates, virtually all simian primates (excluding *Aotus*), and most Lemurinae and Indridae (excluding *Avahi*).

The classical concept of the ecological niche involves, among other things, various adaptations of a species which have developed as a function of the food resources for which it is specialised. As an additional facet of this qualitative adaptation, it must be remembered that the food resources are generally accessible round the clock. For any given organism, the efficiency with which food can be gathered varies over the course of each 24-hour period. As a rule, for a particular category of diet, there are two different mammal species (or species from other groups) which feed in succession – one at night and the other by day. Thus, the activity rhythm (nocturnal or diurnal) can act as a mechanism ensuring ecological separation of certain species whose dietary requirements are identical. In ecological terms, the 'diurnal world' and the 'nocturnal world' are utterly different; they require quite different means of communication and techniques of hunting or location of plant food (predominance of olfaction in nocturnal species and emphasis on vision in diurnal species). This distinction is of considerable importance, both for the study of the evolution of social organisation and for the interpretation of many biological features which are closely associated with the activity rhythm.

In the ecosystem represented by the equatorial rainforest of Gabon, surveys to date have identified 120 mammal species and at least 216 bird species, all of which are sympatric and occupy different ecological niches in the overall system. Among the mammals, 70% are strictly nocturnal, 10% are active both at night and during the day, and only 20% are strictly diurnal. With the birds, on the other hand, 96% of the species are diurnal. A comparable situation is seen in other tropical forest ecosystems, and as an overall observation it emerges that the birds generally occupy the diurnal niches whilst the mammals occupy nocturnal ecological niches. Certain taxonomic groups of mammals (such as monkeys and squirrels), as a result of specialisations, have been able to occupy given diurnal niches in which they appear to be better adapted than the birds. With the monkeys, for example, such specialisations include the evolution of an advanced nervous system, modification of the hand and of its nervous control, and (usually) increase in body size. As a general rule, these three factors are associated, but large body size is not found in all cases (witness the marmosets and the tamarins).

Advanced evolution of the central nervous system and the sophisticated coordination of the hand are intimately linked, and together they represent the most striking feature of the 'higher' (simian) primates. The monkeys, many of which are consumers of animal protein, can forage in dense vegetation, underneath bark and in leaf-litter, using their mobile and highly coordinated hands. The various foraging techniques involve both great dexterity and 'intelligent' behaviour in order to approach or discover animal prey. Some of the simians (colobus monkeys, langurs, howler monkeys, siamangs, gorillas) obtain the proteins they require from green, tough vegetation, which they are able to triturate with the large dentitions associated with their relatively large body-sizes (Hladik et al., 1971). In fact, there is a range of intermediates between these two extremes, but it is generally accepted that folivorous simians derived from ancestors with omnivorous-insectivorous habits. Hence, the evolution of the hand and intelligence of simian primates can be considered as a 'strategy' which permits this mammalian group to occupy certain diurnal ecological niches.

At night, techniques for location of insects are naturally different, because of the low residual light intensity produced by the filtering effect of the foliage. Under nocturnal conditions, the bats present the most obvious parallel to the diurnal birds, and they constitute the majority of the small-bodied predators. Small-bodied, arboreal climbing mammals play a secondary role, but they are represented in virtually all of the tropical, equatorial forest ecosystems. In Africa, the forms involved are the Galaginae and the lorisines; in Asia, there are the tarsiers and/or the lorisines; in Madagascar, there are the Cheirogaleinae; in South America, there are the didelphid marsupials; and in Australia there are the phalangerid marsupials. These small mammals, whose weight is generally less than 1 kg, eat fruits, gums and insects. With the prosimians, at least, the insect prey are trapped directly, either whilst resting or in flight, without any foraging in the vegetation. The fingers of the bushbabies and of the mouse and dwarf lemurs (cheirogaleines) are long and slender, and it is the terminal phalange of each digit – flattened and equipped with longitudinal dermatoglyphs – which plays the primary role in prehension. Insect prey are seized with a rapid, stereotyped movement of one or both hands, in many cases when taking off. Bishop (1964) has demonstrated that many prosimians (lemurs and lorises) have only quite weak neuromuscular control of the hand and that they are unable to carry out separate movements of their digits as seen with the simian primates. It can be concluded that the morphology and neuromuscular control of the prosimian hand has evolved as a function of a form of nocturnal hunting behaviour in which the prey are captured during their activity phase, when they are mobile and/or ready to take flight.

Just as the capture techniques have evolved differently according to whether hunting occurs during the daytime or at night, the sense-organs have evolved in two different directions in the simians and in the prosimians. Among the simians, the good daytime light conditions have encouraged the evolution of sophisticated visual systems: binocular/stereoscopic vision, good colour discrimination, high visual acuity and development of a fovea in the retina. Such good vision is of great value in the detection of uniformly coloured, cryptic prey which are difficult to discern when hidden in vegetation. In addition, fruits can be detected at a distance on the basis of their distinctive colours among the foliage. The nocturnal prosimians, on the other hand, detect their food in many cases by olfactory means. Over and above this, many species which engage in active hunting of insects rely particularly on their fine hearing for prey-localisation. Well-developed ear pinnae which can be oriented over a wide range of directions permit detection of the very faint sounds produced by insects (see p.38). The rapidity and precision of the leaps which the bushbabies and certain lemurs can perform at night in the forest environment, where the canopy allows only one hundredth of the light to filter through, are a good indication of the degree to which the visual apparatus of the nocturnal lemurs has become specialised for conditions of low light intensity. This specialisation, which consists of hypertrophy of the eyeball, the development of a reflecting tapetum behind the retina, and the absence of cones and a fovea (resulting in the absence of colour vision), has occurred in a direction diametrically opposed to that followed by the simian visual system. It would appear that vision in the prosimians does play a major part in the perception of the environment, but that in the search for food the primary role is played by olfaction and hearing.

The few diurnal lemurs pose a problem. Apart from their large body size, they are – in terms of the structure of the hand, the moderate elaboration of the central nervous system and the development of the eye – strikingly similar to the nocturnal prosimians. The hand, constructed on the same plan as that of the Galaginae or the Cheirogaleinae, is incapable of complex manipulation. Further, no diurnal Madagascar lemur occupies an ecological niche directly comparable to that of the typical frugivorous-insectivorous simian of the New or Old Worlds, which catches its insect prey by foraging in the vegetation. The diurnal lemurs all obtain the proteins they require from seeds and green vegetation, as do colobus monkeys, langurs or howler monkeys. The eye in these diurnal lemurs is basically of the nocturnal type, slightly modified for conditions of high light intensities. In almost all species there is a reflecting tapetum containing riboflavin crystals, which is masked to varying degrees by pigment granules (Pariente, 1970; Alfieri et al., 1974). A fovea is scarcely developed or nonexistent, and although colour vision has been demonstrated in a number of species, there is no real comparison with the powers of colour discrimination. exhibited by the diurnal simians. These distinctive characteristics, associated with the fact that many of the diurnal lemurs exhibit a certain propensity for crepuscular activity (Petter, 1962) – or even actual nocturnal activity on some occasions – supports the view that these prosimians are derived from a clearly nocturnal ancestor. The unusual ecological 'vacancies' existing on Madagascar, because of its isolation, have permitted adaptive radiation from a lemur ancestor, already too specialised for nocturnal life to allow the few diurnal offshoots to develop a detailed resemblance to the simians of the New and Old Worlds. It is true that studies based on the comparative anatomy of the tympanic bulla and its carotid circulation indicate similarities between the Eocene Adapidae and the diurnal lemurs of Madagascar which are not shared by the nocturnal lemurs and the Lorisidae (Szalay and Katz, 1973; Szalay, 1975; Hoffstetter, 1974; Cartmill, 1975). However, the fact that there is considerable variation in cranial anatomy (particularly in the tympanic bulla and carotid circulation) within the Madagascar lemurs – with a clear association with body size – may well indicate that adaptive characters linked to specific ecological conditions are involved. Unfortunately, skulls are only known for the larger Adapidae, and it is perfectly possible that this Eocene family exhibited the same range of variation in cranial characters. The cranial anatomy of the larger bodied, diurnal Madagascar lemurs, even if it shows resemblances to that of the larger Adapidae, need not necessarily represent a primitive stage of evolution. The condition shared by the nocturnal lemurs and lorises could equally well be primitive for the lemurs and lorises generally. The considerable gaps in the fossil record leave a large element of doubt regarding the evolution of the existing lemurs and lorises; but the ecological arguments developed above provide strong evidence that the group as a whole should be regarded as highly specialised for nocturnal life.

If one attempts to 'reconstitute' a hypothetical common ancestor for the living primates, it must at once be recognised that this ancestor must have possessed eyes which were not particularly specialised either for nocturnal vision or for diurnal vision, and were hence capable of modification to the condition found in modern prosimians and to the condition typical of the simians. By the same token, this hypothetical ancestor must have had hands of a generalised conformation, capable of giving rise to either the prosimian type or the simian type in evolution. Most of the diurnal and nocturnal ecological niches currently occupied by modern primates must have been already occupied by Palaeocene and Eocene prosimians. It is perfectly possible that, in the early stages of primate evolution, certain relatively unspecialised lineages may have passed alternately through nocturnal and diurnal stages. Nevertheless, the dichotomy between diurnal and nocturnal habits must have isolated at quite an early stage the two branches of the primates, which subsequently built upon their evolutionary tendencies in a virtually irreversible fashion.

The existing dichotomy between strepsirhine and haplorhine primates must have originated a considerable time ago, and the modern lemurs and lorises are probably derived from an ancestral strepsirhine which was already clearly separated from the lineage which eventually gave rise to the modern simians. This point of view is supported by numerous publications in other fields, covering morphogenesis of the placenta and the foetal membranes (Luckett, 1974, 1975), the conformation of the central nervous system and the sense organs (Martin, 1973), general reproductive characters (Martin, 1975) and biochemical comparisons (Goodman, 1974, 1975).

The Order Primates, viewed in terms of overall evolutionary tendencies, stands out above all as a group specialised in terms of central nervous developments. The most advanced representatives (the simians) are essentially characterised by adaptation for diurnal ecological niches in association with further elaboration of the nervous system and the sense organs. Diurnal life has opened to this group of primates a number of evolutionary possibilities, such as increased complexity of social structure (almost always on the basis of gregarious habits) and refinement of cerebral processes, associated with an increase in body size and a number of secondary adaptations which have permitted occupation of a fairly wide range of diurnal ecological niches.

In contrast with these trends, it would seem that nocturnal life has exerted a more 'conservative' influence, judging by the large number of primitive characters preserved among the surviving prosimians. Nevertheless, selection pressures leading towards specialisation have continued to operate even with this group of primates, though the characters affected are different from those which have been influenced in the evolution of the 'higher' primates. It can be postulated that with the strepsirhine primates (lemurs and lorises), development of intelligence and of the nervous system on which it is based ceased to evolve in the direction typical of the diurnal primates from the point where these animals had become too specialised for nocturnal life. Thus, the modern lemurs represent a marginal group which has generally maintained primitive characters of the order Primates at the time of its origin. In this respect, the strepsirhine primates are living witnesses of great value in the attempt to interpret the early evolution of the primates. However, great care must be exercised in the process of identifying primitive characteristics; a distinction must be drawn between 'primitive primate characters' and 'characters of nocturnal prosimians'.

With this view of the broad evolutionary framework within which the Lorisidae of Africa and Asia eventually made their appearance, a more detailed interpretation is possible of the species covered in the present study. The five lorisid species studied in Gabon are components of an extremely complex ecosystem. The relative climatic stability of the equatorial rainforest, together with the diversity of available biotopes, has favoured the development of a multiplicity of ecological niches and thus the emergence of a wide range of plant and animal species. Each animal species exploits a precisely determined spectrum of dietary resources in a well-defined biotope, and it was therefore possible to establish the ecological parameters upon which each lorisid species depends. The level of the forest exploited, the nature of the supports used, and the food resources which are sought after, are all associated with morphological and behavioural adaptations which would remain enigmatic without a detailed knowledge of the ecological peculiarities of each species. The claw-like extensions of the nails in *Euoticus elegantulus*, the notable elongation of its tooth-scraper and its tongue, the enlargement of its caecum, the enlargement and flattening of its terminal digital pads, and its 'style' of leaping (permitting landing with all four limbs simultaneously on vertical or oblique trunks), are all explicable when it is known that this species seeks out gums, particularly on large-diameter trunks and branches with smooth surfaces. Similarly, the peculiar leaping 'style' of *Galago alleni*, involving rotation around thin vertical supports achieved through its kinetic energy and a whip-like action of the tail, would seem to be clearly adapted to movement in undergrowth areas where only small-diameter trunks and vertical liane bases are available. The small body size of *Galago demidovii*, and its versatile locomotor repertoire, are correlated with its movement in dense vegetation containing inter-twined branches and lianes, where it seeks out its insect prey. Finally, the distinctive physiques of *Perodicticus potto* (heavy and robust) and of *Arctocebus calabarensis* (light and slender) are associated with the divergent ecological adaptations of these two species.

In addition to these adaptations, directly or indirectly associated with feeding behaviour, defensive behaviour towards predators has exerted a profound influence on the locomotor patterns of the various lorisid species. Two basic 'strategies' have emerged: (1) fleeing by means of rapid running and/or leaping (Galaginae); (2) unobtrusive, slow and silent climbing, to avoid drawing the attention of predators (Lorisinae). In anatomical terms, these two different subfamilies are distinguished primarily by the morphological structure of their limbs, and it is difficult to know how far back in time the separation between the two lineages can be traced. The Miocene fossil lorisids, dating back to about 20 million years ago, are generally quite similar to modern bushbabies in terms of the characters which have been preserved, but the elongation of the tarsal bones of the foot is less pronounced and closer to that found in the extant Cheirogaleinae of Madagascar (Walker, 1970; Martin, 1972b). These Miocene tarsal bones are generally attributed to species of the genus *Progalago*, but Walker (1974) allocates skull and jaw fragments of another East African Miocene lorisid, *Mioeuoticus* (2 species), to the subfamily Lorisinae, on the basis of the prominent ridges on the skull roof, which are also seen in all modern lorisine species. Thus, it is possible that the Galaginae were already distinct from the Lorisinae by the Miocene.

The selection pressures exerted by predation have doubtless existed continuously throughout the evolution of the lorisids, and at the same time the various small animal prey species have probably been evolving more elaborate protective mechanisms as the lorisids have been evolving more refined hunting techniques to counter the protection thus afforded.

Although the rapid flight capacity of the Galaginae has often been considered as an effective mechanism for escape from predators, the same cannot be said of the slow-climbing habit of the Lorisinae, which has been regarded by several authors as no more than a 'bizarre product' of nature. In fact, arboreal predators locate their prey primarily on the basis of sound and detection of sudden movements made in the foliage. Thus, the pattern of movement of the Lorisinae can be regarded as a form of 'cryptic locomotion', which is only of value in dense forest where the screening effect of the foliage and fine branches acts to camouflage silhouettes. As expected, the Lorisinae are not found in open forest areas in which bushbabies can still be found. In order to operate more efficiently as an anti-predator mechanism, this slow-climbing pattern of the lorisines is accompanied by complex neuromuscular coordination permitting movement without any abrupt transitions. In addition, the hands of the lorisines have been considerably modified to become powerful pincers, with reduction of the length of the phalanges and atrophy of the index finger. In its turn, this strategy of dissimulation rather than active flight has had its effects on the selection of dietary components and on social behaviour; discretion in movement is matched by economy. The lorisines are specialised for capture of irritant or noxious prey, which are relatively easy to detect and capture and are ignored by most other insect predators. Social relationships have become discreet in the extreme; there are no powerful vocalisations, and communication is largely based on olfactory messages. In this way, actual encounters between neighbours (and hence the distances required in locomotion) are reduced to a strict minimum. In all of these features, the lorisines contrast markedly with the galagines, which are actively-moving, highly vocal forms with more obvious social contacts. Overall, it seems improbable that specialisations for running and/or jumping would have exerted such profound effects on the general biology of the Galaginae, and hence it is to this subfamily that we should look in order to seek information for reconstituting the common ancestor of the strepsirhine primates.

Taking into account all the available evidence – morphological, behavioural and ecological – it seems likely that the common ancestor of the strepsirhines would have been small to medium in size, within the range now covered by the Galaginae and the Cheirogaleinae (Martin, 1972b), and that its adaptations centred upon a nocturnal habit. The following list of behavioural characters can be compiled, in accordance with previous publications (Charles-Dominique and Martin, 1970) and with the results of the study conducted in Gabon:

1. Climbing and leaping habit, with the hind limbs and the constituent tarsal bones less elongated than in the specialised modern offshoots.
2. Diet consisting of gums, fruit and small arthropods.
3. Construction of spherical leaf-nests.
4. Offspring born at a relatively advanced stage compared to modern insectivores, but at a less advanced stage than most modern primates.
5. Oral transport of the infant, with the flank grasped in the mother's teeth.
6. Social communication based partially on urine-marking.
7. Solitary mode of life.
8. Overlap between male and female home ranges.
9. Dispersal of young males.
10. Continued residence of maturing females, with persistence for some time of the relationship with the mother.

These characters, which seem likely to have been present in the ancestral strepsirhines, are found in several other mammalian orders, and many of them probably represent characteristics of the 'primitive placental mammal', which would have given rise to the ancestral primates and hence to the dominant lineages of the 'higher' primates.

In addition to these apparently primitive characters, one finds a number of specialisations which are restricted to the strepsirhine primates:

1. Eye possessing a reflecting tapetum containing riboflavin crystals, and adapted for vision at low light intensities.
2. Hands adapted for the capture of small arthropod prey in flight or at take-off; digits elongated, retroflexion of the terminal phalanges, and large digital pads on the ventral surfaces of the terminal phalanges.

3. Possession of a tooth-scraper in the lower jaw, associated with cornification of the retained sublingua. Although it is certainly true that the tooth-scraper is widely used for grooming in strepsirhines (Buettner-Janusch and Andrew, 1962), its evolutionary origin has been linked by Martin (1972b) to its use as a scoop for the collection of gums and other dietary items which are not readily accessible for ingestion.

These specialisations are relatively minor, and the adaptation of the hands is shared with the tarsiers, but they are sufficient to permit this basically primitive group of primates to survive in ecological niches available for small-bodied, nocturnal mammals, subsisting on a mixed diet of insects, gums and fruits.

It is, of course, highly speculative to attempt to reconstruct the characteristics of the ancestral strepsirhines largely on the basis of behavioural and ecological considerations. More detailed reconstruction of strepsirhine evolution depends upon an effective synthesis of information from a wide range of biological disciplines – including palaeontology and comparative biochemistry. However, it has clearly emerged from this study of five sympatric lorisid species in Gabon that detailed examination of behavioural and ecological relationships can yield significant new information for the interpretation of the anatomical characters of living species and for interpreting past evolutionary trends. In particular, analysis of the distinctions between these sympatric, nocturnal primates has shown how specialisations can arise in connection with adaptation to distinct ecological niches within the same forest ecosystem.

Bibliography

Alfieri, R., Pariente, G and Sole, P. (1974). Dynamic electroretinography in monochromatic light and fluorescence electroretinography in lemurs. *12th Int. Sympos. Soc. Clin. E.R.G.* 1974.

Andrew, R.J. (1963). The origin and evolution of the calls and facial expressions of the Primates. *Behaviour* 20, 1-109.

Andrew, R.J. (1964). The displays of the Primates. In *Evolutionary and Genetic Biology of Primates*, Vol. 2 (ed. Buettner-Janusch, J.). Academic Press, New York.

Andrew, R.J. and Klopman, R.B. (1974). Urine washing: comparative notes. In *Prosimian Biology* (ed. Martin, R.D., Doyle, G.A. and Walker, A.C.). Duckworth, London.

Bates, G.L. (1905). Notes on the Mammals of Southern Cameroons and the Benito. *Proc. Zool. Soc. Lond.* 1905, 65-85.

Beaudenon, P. (1949). Contribution à la connaissance du Potto de Bosman, Togo Sud. *Mammalia* 13, 76-99.

Bearder, S.K. and Doyle, G.A. (1974). Ecology of bushbabies, *Galago senegalensis* and *Galago crassicaudatus*, with some notes on their behaviour in the field. In *Prosimian Biology* (ed. Martin, R.D., Doyle, G.A. and Walker, A.C.). Duckworth, London.

Beerten-Joly, A., Piavaux, A. and Goffart, M. (1974). Quelques enzymes digestives chez un prosimien, *Perodicticus potto*. *C.R. Acad.Soc.Biol.* 168, 140.

Bishop, A. (1964). Use of the hand in lower Primates. In *Evolutionary and Genetic Biology of Primates*, Vol. 2. (ed. Buettner-Janusch, J.). Academic Press, New York and London.

Blackwell, K.F. and Menzies, J.L. (1968). Observations on the biology of the potto (*Perodicticus potto*, Müller). *Mammalia* 32, 447-51.

Boulenger, E.G. (1936). *Apes and Monkeys*. Harrap, London.

Brosset, A. (1968). Localisation écologique des oiseaux migrateurs dans la forêt équatoriale du Gabon. *Biol. Gabon.* 4, 211-26.

Brosset, A. (1971). Recherches sur la biologie des Pycnonotidés du Gabon. *Biol. Gabon.* 7, 423-60.

Buettner-Janusch, J. (1964). The breeding of galagos in captivity and some notes on their behavior. *Folia primat.* 2, 93-110.

Buettner-Janusch, J. and Andrew, R.J. (1962). The use of the incisors by primates in grooming. *Amer. J. Phys. Anthrop.* 20, 127-9.

Butler, H. (1957). The breeding cycle of the Senegal Galago, *Galago senegalensis senegalensis*, in the Sudan. *Proc. Zool. Soc. Lond.* 129, 147-9.

Butler, H. (1960). Some notes on the breeding cycle of the Senegal Galago, *Galago senegalensis senegalensis*, in the Sudan. *Proc. Zool. Soc. Lond.* 135, 423-30.

Butler, H. (1967). Seasonal breeding of the Senegal Galago (*Galago senegalensis senegalensis*) in the Nuba mountains, Republic of the Sudan. *Folia Primat.* 5, 165-75.

Cansdale, G.S. (1946). The lesser bushbaby *Galago demidovii demidovii* G. Fish. *J. Soc. Pres. Fauna Empire, Lond.* 50, 6-12.

Cansdale, G. (1960). *Bush Baby Book.* Phoenix House, London.

Cartmill, M. (1975). Strepsirhine basicranial morphology and the affinities of the Lorisiformes. In *Phylogeny of the Primates: a multidisciplinary approach* (ed. Luckett, W.P. and Szalay, F.S.). Plenum Press, New York.

Chappuis, C. (1971). Un exemple de l'influence du milieu sur les émissions vocales des oiseaux: l'évolution des chants en forêt équatoriale. *Terre et Vie*, 25, 183-202.

Charles-Dominique, P. (1966a). Naissance et croissance d'*Arctocebus calabarensis* en captivité. *Biol. Gabon.* 2, 331-45.

Charles-Dominique, P. (1966b). Glandes préclitoridiennes de *Perodicticus potto*. *Biol. Gabon.* 2, 355-359.

Charles-Dominique, P. (1968). Reproduction des lorisidés africains. In *Cycles génitaux saisonniers de Mammifères sauvages. Entretien de Chizé, sér. Physiol.* 1, 2-5. Masson, Paris.

Charles-Dominique, P. (1971a). Eco-éthologie et vie sociale des prosimiens du Gabon. *Thèse de Doctorat d'Etat*, Paris (C.N.R.S. No. A.O. 5816).

Charles-Dominique, P. (1971b). Eco-éthologie des prosimiens du Gabon. *Biol. Gabon.* 7, 121-228.

Charles-Dominique, P. (1972). Ecologie et vie sociale de *Galago demidovii* (Fischer 1808, Prosimii). *Z.f. Tierpsychol.* suppl. 9, 7-41.

Charles-Dominique, P. (1974a). Aggression and territoriality in nocturnal prosimians. In *Primate Aggression, Territoriality and Xenophobia* (ed. Holloway, R.L.). Academic Press, New York.

Charles-Dominique, P. (1975a). Vie sociale de *Perodicticus potto* (Primate, Lorisidés). Etude de terrain en forêt équatoriale de l'Ouest africain au Gabon. *Mammalia* 38, 355-79.

Charles-Dominique, P. (1975b). Nocturnal primates and diurnal primates: an ecological interpretation of these two modes of life by analysis of the higher vertebrate fauna in tropical forest ecosystems. In *Phylogeny of the Primates: a multidisciplinary approach* (ed. Luckett, W.P. and Szalay, F.S.). Plenum Press, New York.

Charles-Dominique, P. and Hladik, C.M. (1971). Le Lépilémur du Sud de Madagascar: écologie, alimentation et vie sociale. *Terre et Vie* 25, 3-66.

Charles-Dominique, P. and Martin, R.D. (1970). Evolution of lorises and lemurs. *Nature* 227, 257-60.

Cowgill, U. (1969). Some observations on the prosimian *Perodicticus potto*. *Folia primat.* 2, 144-150.

Cowgill, U.M. (1974). Co-operative behaviour in *Perodicticus potto*. In *Prosimian Biology* (ed. Martin, R.D., Doyle, G.A. and Walker, A.C.). Duckworth, London.

Crook, J.H. and Gartlan, J.S. (1966). Evolution of primate societies. *Nature, Lond.* 210, 1200-3.

Doyle, G.A., Andersson, A. and Bearder, S.K. (1971). Reproduction in the Lesser Bushbaby (*Galago senegalensis moholi*) under semi-natural conditions. *Folia primat* 14, 15-22.

Doyle, G.A. and Bekker, T. (1967). A facility for naturalistic studies of the lesser bushbaby (*Galago senegalensis moholi*). *Folia primat.* 7, 161-8.

Doyle, G.A., Pelletier, A. and Bekker, T. (1967). Courtship, mating and parturition in the lesser bushbaby (*Galago senegalensis moholi*) under seminatural conditions. *Folia primat.* 7, 169-97.

Dubost, G. (1968). Le rythme annuel de reproduction du chevrotain aquatique, *Hyemoschus aquaticus* Ogilby, dans le secteur forestier du Nord-Est du Gabon. In *Cycles genitaux saisonniers de Mammifères sauvages. Entretien de Chizé, ser. Physiol.* 1, 51-65. Masson, Paris.

Eibl-Eibesfeldt, I. (1953). Eine besondere Form des Duftmarkierens beim Riesengalago, *Galago crassicaudatus* E. Geoffroy, 1812. *Saugetierk. Mittl.* 1, 1713.

Eimerl, S. and DeVore, I. (1965). *The Primates.* Time-Life, New York.

Eisenberg, J.F. and Gould, E. (1970). The tenrecs: a study in mammalian behavior and evolution. *Smithson. Contrib. Zool.* 27, 1-137.

Epps, J. (1974). Social interactions of *Perodicticus potto* kept in captivity in Kampala, Uganda. In *Prosimian Biology* (ed. Martin, R.D., Doyle, G.A. and Walker, A.C.). Duckworth, London.

Fogden, M.P.L. (1974). A preliminary field study of the western tarsier, *Tarsius bancanus* Horsefield. In *Prosimian Biology* (ed. Martin, R.D., Doyle, G.A. and Walker, A.C.). Duckworth, London.

Geoffroy St. Hilaire, E. (1812). Tableau des quadrumanes. *Ann. Mus. Hist. Nat. Paris.* 19, 157-70.

Goffart, M. and Hildwein, G. (1972). Défense métabolique chez un Prosimien, *Perodicticus potto*. *C.R. S. Soc. Biol.* 166, 1382.

Goodman, M. (1975). Protein sequence and immunological specificity in the phylogenetic study of primates. In *Phylogeny of the Primates: a multidisciplinary approach* (ed. Luckett, W.P. and Szalay, F.S.). Plenum Press, New York.

Grassé, P.P. (1975). *Traité de Zoologie* 17: Les Mammifères, 1642-53.

Gregory, W.K. (1915). The classification and the phylogeny of the Lemuroidea. *Bull. Geol. Soc. Amer.* 26, 419-46.

Haddow, A.J. and Ellice, J.M. (1964). Studies on bushbabies with special reference to the-epidemiology of yellow fever. *Trans. Roy. Soc. trop. Med. Hyg.* 58, 521-38.

Hall-Craggs, E.C.B. (1965). An analysis of the jump of the lesser galago (*Galago senegalensis*). *J. Zool.* 147, 20-9.

Hall-Craggs, E.C.B. (1974). Physiological and histochemical parameters in comparative locomotor studies. In *Prosimian Biology* (ed. Martin, R.D., Doyle, G.A. and Walker, A.C.), Duckworth, London.

Hayman, R.W. (1937). A note on *Galago senegalensis inustus* Schwarz. *Ann. Mag. Nat. Hist.* 20, 149-51.

Hill, W.C.O. (1938). A curious habit common to lorisoids and platyrrhine monkeys. *Ceylon J. Sci. (B)* 21, 66.

Hill, W.C.O. (1953). *Primates, Comparative Anatomy and Taxonomy*, Vol. 1: *Strepsirhini.* University Press, Edinburgh.

Hladik, A. and Hladik, C.M. (1969). Rapports trophiques entre végétation et primates dans la forêt de Barro Colorado (Panama). *Terre et Vie.* 23, 25-117.

Hladik, C.M., Hladik, A., Bousset, J., Valdebouze, P., Viroben, G., and Delort-Laval, J. (1971). Le régime alimentaire des primates de l'île de Barro Colorado (Panama); résultats des analyses quantitatives. *Folia primat.* 16, 85-122.

Hoffstetter, R. (1974). Phylogeny and geographical deployment of the Primates. *J. Hum. Evol.* 3, 327-50.

Ilse, D.R. (1955). Olfactory marking of territory in two young male loris, *Loris tardigradus lydekkerianus*, kept in captivity at Poona. *Brit. J. Anim. Behav.* 3, 118-20.

Ioannou, J.M. (1966). The oestrous cycle of the Potto. *J. Reprod. Fertil.* 11, 455-7.

Jarvis C. and Morris, D. (1960). Longevity survey: length of life of mammals in captivity at the London Zoo and Whipsnade park. *Int. Zoo. Yb.* 2, 288-99.

Jewell, P.A. and Oates, J.F. (1969a). Breeding activity in prosimians and small rodents in West Africa. *J. Reprod. Fertil. Suppl.* 6, 23-38.

Jewell, P.A. and Oates, J.F. (1969b). Ecological observations on the lorisoid primates of African lowland forest. *Zool. africana* 4, 231-48.

Jolly, A. (1966). *Lemur behaviour: a Madagascar field study.* University Press, Chicago.

Jones, C. (1969). Notes on ecological relationship of four species of lorisids in Rio Muni, West Africa. *Folia primat.* 11, 255-67.

Jouffroy, F.K., Gasc, J.P. and Decombas, M. (1974). Biomechanics of vertical leaping from the ground in *Galago alleni*: a cineradiographic analysis. In *Prosimian Biology* (ed. Martin, R.D., Doyle, G.A. and Walker, A.C.). Duckworth, London.

Jouffroy, F.K. and Gasc, J.P. (1974). A cineradiographic analysis of jumping in an african prosimian (*Galago alleni*). In *Advances in Primatology; 3. Primate locomotion* (ed. Jenkins, F.A.). New York.

Kingdon, J. (1971). *An Atlas of East African Mammals*, Vol. 1. Academic Press, London.

Le Gros Clark, W.E. (1971). *The Antecedents of Man.* University Press, Edinburgh.

Lowther, F. de (1940). A study of the activities of a pair of *Galago senegalensis moholi* in captivity, including the birth and post natal development of twins. *Zoologica, N.Y.* 25, 433-62.

Luckett, W.P. (1974). The phylogenetic relationships of the prosimian primates: evidence from morphogenesis of the placenta and foetal membranes. In *Prosimian Biology.* (ed. Martin, R.D., Doyle, G.A. and Walker, A.C.). Duckworth, London.

Luckett, W.P. (1975). Ontogeny of primate fetal membranes and placenta and their bearing on mammalian and primate phylogeny. In *Phylogeny of* the *Primates: a multidisciplinary approach* (ed. Luckett, W.P. and Szalay, F.S.). Plenum Press, New York.

Machida, H. and Giacometti, L. (1967). The anatomical and histochemical properties on the skin of the external genitalia of the Primates. *Folia primat.* 6, 48-69.

Malbrant, R. and Maclatchy, A. (1949). *Faune de l'équateur africain francais. II: Mammifères.* Lechevalier, Paris.

Manley, G.H. (1966). Reproduction in lorisoid primates. *Symp. Zool. Soc. Lond.* 15, 493-509.

Manley, G.H. (1967). Gestation periods in the Lorisidae. *Int. Zoo. Yb.* 7, 80-1.

Manley, G.H. (1974). Functions of the external glands of *Perodicticus* and *Arctocebus.* In *Prosimian Biology* (ed. Martin, R.D., Doyle, G.A. and Walker, A.C.). Duckworth, London.

Martin, R.D. (1972a). A preliminary field study of the lesser mouse lemur (*Microcebus murinus* J.F. Miller 1777). *Z.f. Tierpsychol. Suppl.* 9, 43-89.

Martin, R.D. (1972b). Adaptative radiation and behaviour of the Malagasy Lemurs. *Phil. Trans. Roy. Soc. Lond. (B)* 264, 295-352.

Martin, R.D. (1973a). A review of the behaviour and ecology of the lesser mouse lemur (*Microcebus murinus*, J.F. Miller 1777). In *Comparative Ecology and Behaviour of Primates* (ed. Michael, R.P. and Crook, J.H.). Academic Press, London.

Martin, R.D. (1973b). Comparative anatomy and primate systematics. *Symp. Zool. Soc. Lond.* 33, 301-37.

Martin, R.D. (1975). The bearing of reproductive behaviour and ontogeny on strepsirhine phylogeny. In *Phylogeny of the Primates: a multidisciplinary approach* (ed. Luckett, W.P. and Szalay, F.S.). Plenum Press, New York.

Mivart, St. G. (1864). Notes on the crania and dentition of the Lemuridae. *Proc. Zool. Soc. Lond.* 1864, 611-48.

Montagna, W. (1962). The skin of primates. VII: the skin of the great bushbaby (*Galago crassicaudatus*). *Amer. J. Phys. Anthrop.* 20, 149-66.

Montagna, W. and Ellis, R.A. (1959). The skin of Primates. I: The skin of *Perodicticus potto. Amer. J. Phys. Anthrop.* 17, 137-62.

Montagna, W. and Ellis, R.A. (1960). The skin of the Primates. II: The skin of the slender loris (*Loris tardigradus*). *Amer. J. Phys. Anthrop.* 18, 19-44.

Montagna, W. and Lobitz, W. (1963). The mammalian epidermis and its derivatives. *Symposia of the Zoological Society of London* 12, 9738.

Montagna, W., Yasuda, K. and Ellis, R.A. (1961). The skin of Primates. III: The skin of the slow loris (*Nycticebus coucang*). *Amer. J. Phys. Anthrop.* 19, 1-21.

Montagna, W. and Yun, J.S. (1962). Further observations on *Perodicticus potto. Amer. J. Phys. Anthrop.* 20, 441-50.

Montagna, W. and Yun, J.S. (1962). The skin of Primates. X: The skin of the ring tail lemur (*Lemur catta*). *Amer. J. Phys. Anthrop.* 20, 95-118.

Napier, J.R. and Napier, P.H. (1967). *A Handbook of Living Primates.* Academic Press, New York.

Napier, J.R. and Walker, A.C. (1967). Vertical clinging and leaping, a newly recognised category of locomotor behaviour among Primates. *Folia primat.* 6, 180-203.

Odum, E.P. (1953). *Fundamentals of ecology.* W.B. Saunders, Philadelphia.

Pages, E. (1970). Sur l'écologie et les adaptations de l'oryctérope et des pangolins sympatriques du Gabon. *Biol. Gabon.* 6, 27-92.

Pariente, G. (1970). Rétinographies comparées chez les lémuriens malgaches. *G.R. Acad. Sci.* 270, 1404-7.

Pariente, G. (1974). Influence of light on the activity rhythms of 2 Malagasy lemurs: *Phaner furcifer* and *Lepilemur mustelinus leucopus.* In *Prosimian Biology* (ed. Martin, R.D., Doyle, G.A. and Walker, A.C.). Duckworth, London.

Petter, J.-J. (1962). Recherches sur l'écologie et l'éthologie des lémuriens malgaches. *Mém. Mus. nat. Hist. nat. Paris (sér. A)* 27, 1-146.

Petter, J.-J. (in prep.). Behavioural factors and the classification of prosimians. In *The Study of Prosimian Behaviour* (ed. Doyle, G.A. and Martin, R.D.). Academic Press, New York.

Petter, J.-J. and Hladik, C.M. (1970). Observations sur le domaine vital et la densité de population de *Loris tardigradus* dans les forêts de Ceylan. *Mammalia* 3, 394-409.

Petter, J.-J., Schilling, A. and Pariente, G. (1971). Observations éco-éthologiques sur deux lémuriens malgaches nocturnes: *Phaner furcifer* et *Microcebus coquereli. Terre et Vie* 27, 287-327.

Petter-Rousseaux, A. (1962). Recherches sur la biologie de la reproduction des primates inférieurs. *Mammalia* 26 (suppl. 1), 1-88.

Petter-Rousseaux, A. (1968). Cycles genitaux saisonniers des lémuriens malgaches. In *Cycles genitaux saisonniers de Mammiferes sauvages, Entretien de Chizé, sér. Physiol.* I, 11-18. Masson, Paris.

Petter-Rousseaux, A. (1974). Photoperiod, sexual activity and body weight variations of *Microcebus murinus* (Miller, 1777). In *Prosimian Biology* (ed. Martin, R.D., Doyle, G.A. and Walker, A.C.). Duckworth, London.

Pocock, R.I. (1918). On the external characters of the lemurs and of *Tarsius. Proc. Zool. Soc. London* 1918, 19-53.

Rahm, U. (1960). Quelques notes sur le Potto de Bosman. *Bull. IFAN* 23, 331-42.

Ramaswami, L.S. and Anad Kumar, T.C. (1962). Reproductive cycle of the slender loris. *Naturwiss.* 49, 115-16.

Ramaswami, L.S. and Anad Kumar, T.C. (1965). Some aspects of the reproduction of the female slender loris, *Loris tardigradus lydekkerianus,* (Cabr.). *Acta zool.* 46, 257-373.

Richard, G. (1970). Territoire et domaine vital. *Entretien de Chizé, ser. Ecologie et Ethologie,* Masson, Paris.

Romer, A.S. (1966). *Vertebrate Palaeontology.* University Press, Chicago.

Sanderson, I.T. (1937). *Animal Treasure.* MacMillan & Co., London.

Sanderson, I.T. (1940). The mammals of the North Cameroons forest area. *Trans. Zool. Soc. Lond.* 24, 623-725.

Sauer, E.G.F. and Sauer, E.M. (1963). The South West African bushbaby of the *Galago senegalensis* group. *J. S.W. Afr. Sci. Soc.* 16, 5-36.

Seitz, E. (1969). Die Bedeutung geruchlicher Orientierung beim Plumplori *Nycticebus coucang* Boddaert 1785 (Prosimii, Lorisidae). *Z.f. Tierpsychol.* 26, 73-103.

Simpson, G.G. (1945). The principles of classification and a classification of Mammals. *Bull. Amer. Mus. Nat. Hist.* 85, 1-350.

Struhsaker, T.T. (1970). Notes on *Galagoides demidovii* in Cameroon. *Mammalia* 34, 207-11.

Suckling, J.A., Suckling, E.E. and Walker, A.C. (1969). Suggested function of the vascular bundles in the limbs of *Perodicticus potto. Nature* 221, 379-80.

Szalay, F.S. (1975). The origin of primate higher categories: an assessment of the basicranial evidence. In *Phylogeny of the Primates: a multidisciplinary approach* (ed. Luckett, W.P. and Szalay, F.S.). Plenum Press, New York.

Szalay, F.S. and Katz, C.C. (1973). Phylogeny of lemurs, galagos and lorises. *Folia primat.* 19, 88-103.

Tate Regan, C. (1930). The evolution of the Primates. *Ann. Mag. Nat. Hist.* (10th Ser.) 6, 383-92.

Tien, D.V. (1960). Sur une nouvelle espèce de *Nycticebus* du Vietnam. *Zool. Anz.* 164, 240-43.

Treff, H.A. (1967). Tiefensehschärfe beim Galago (*G. senegalensis*). *Z. vgl. Physiol.* 54, 26-57.

Van Kampen, P.N. (1905). Die Tympanalgegend des Säugetierschädels. *Morph. Jahrb.* 34, 321-722.

Vincent, F. (1968). La sociabilité du Galago de Demidoff. *Terre et Vie.* 22, 51-6.

Vincent, F. (1969). Contribution a l'étude des prosimiens africains. Le Galago de Demidoff. *Thèse de Doctorat d'Etat*, Paris. (CNRS No. AO 3575).

Vincent, F. (1972). Prosimiens africains. V: Repartition géographique de *Euoticus inustus. Ann. Fac. Sci. Cameroun.* 10, 135-41.

Walker, A.C. (1967). Locomotor adaptation in recent and fossil Madagascar Lemurs. Ph.D. Thesis. University of London.

Walker, A.C. (1968). A note on the 'spines' on the neck of the Potto. *Uganda J.* 32, 221-2.

Walker, A.C. (1969). The locomotion of the lorises, with special reference to the potto. *E. Afri. Wildl. J.* 7, 1-5.

Walker, A.C. (1970a). Nuchal adaptations in *Perodicticus potto. Primates* 11, 135-44.

Walker, A.C. (1970b). Postcranial remains of the Miocene Lorisidae of East Africa. *Am. J. Phys. Anthrop.* 33, 249-61.

Walker, A.C. (1974). A review of the Miocene Lorisidae of East Africa. In *Prosimian Biology* (ed. Martin, R.D., Doyle, G.A. and Walker, A.C.). Duckworth, London.

Weber, M. (1928). *Die Säugetiere*, Vol. 2: *Systematischer Teil*, Fischer Verlag, Stuttgart.

Index of Authors

Figures in bold refer to illustrations

General Index

Figures in bold refer to illustrations

General Index 277

Diospyros piscatoria, 45; Diospyros
hoyeleana, 45; Entada gigas, 41, 42,
43; Mimosaceae, 41; Musanga
cecropioides (parasol tree), 22, **23**,
45; Pentacletra eetveldeana, 41, 46;
Piptadenastrum africanum, 41;
Polyaltia suaveoleus, 45;
Ricinodendron africanus, 45; tree-
fall zone, 60, 62, 66, 79, 128, 129,
131
trunk, 20, 42, 63, 65, 109, 111,
hollow, 119, 125, 189, **209**, 227,
242
Tupaiidae (tree shrews), 248;
Ptilocercus, 249

Uapaca, 45, 46, 47
urine, 77, 195, 197, 206, 216;
deposition, 77; marking, 195,
199, 200, 206, 212, 213, 219, 222,
226, 258; trail, 197; washing,
196, 197; and see marking,
olfactory communication

vegetation; canopy, 27, 31, **32**, 40,
59, 65, 75, 77, 84, 112, 127, 252;
dense, 62, 63, 65, 66, 87, 107,
124; foliage, 55; rustling, 69, 96;
strata, 20, **56**; stratification, 40;
undergrowth, 20, **25**, 31, **32**, 40,
46, 63, **64**, 79, 96, 127, 129, 176,
186, 187, 188, 256

'visibility profile', see observation
vision, 1, 55; and see eye
visual signals, 164; gesture, threat,
165; hands, 164; and see posture
viverrids; arboreal, 192; Atilax
paludinosus (mongoose), 83;
Bdeogale nigripes (black-legged
mongoose), 84, 85; Civettictis
civetta, 83; Genetta servalina, 83,
92; Genetta tigrina, 83; Poiana
richardsoni (African linsang), 83,
84; Nandinia binotata (African
palm civet), 83, 84, 92, 93, 115,
150
vocalisation, 54, 63, 66, 85, 93, 94,
146, 164, 195, 212, 219, 226, 230,
245, 257; aggressive, 94, 218,
246; agonistic, **172**; function,
185; growling, 188;
individualised, 169, 181; long-
range, 189; vocal signals, 195;
and see calls, sonagram

warbler (Camaroptera brevicaudata),
see birds
weaning, 49, 52
weaver-bird (Ploceus cuculatus), see
birds
weights; body, 15, 78, 149, 161, 164;
birth, **140**; embryo, 143;
increase, 162